Transport and Energy Processes 2016

Core Programming Area at the 2016 AIChE Annual Meeting

San Francisco, California, USA
13-18 November 2016

ISBN: 978-1-5108-3450-7

Printed from e-media with permission by:

Curran Associates, Inc.
57 Morehouse Lane
Red Hook, NY 12571

Some format issues inherent in the e-media version may also appear in this print version.

Printed by Curran Associates, Inc. (2017)

For permission requests, please contact AIChE
at the address below.

AIChE
120 Wall Street, FL 23
New York, NY 10005-4020

Phone: (800) 242-4363
Fax: (203) 775-5177

www.aiche.org

Additional copies of this publication are available from:

Curran Associates, Inc.
57 Morehouse Lane
Red Hook, NY 12571 USA
Phone: 845-758-0400
Fax: 845-758-2633
Email: curran@proceedings.com
Web: www.proceedings.com

TABLE OF CONTENTS

472746 Thermodynamic Properties and Transport of Fluids in Presence of Cross-Linked Polymers and Confinement: Modeling, Simulations and Experiments

Abstract:

Research Interests: Thermodynamics, Fluid Dynamics, Nanofluidics, Fuels, Molecular Simulations, Oil and Gas, CO2 sequestration

Teaching Interests: Thermodynamics, Fluid Dynamics, Reservoir Characterization and Engineering, Molecular Simulations

Thermodynamic properties of fluids change in presence of cross linked polymers and in confined spaces. Correct prediction of deviation in these properties is valuable to many industries including the oil and gas industry. In last decade, the shale play resources have played a key role in increasing oil production in the United States. However, the average hydrocarbon recovery factors from many of these formations are usually less than 10%. There is a considerable interest in understanding the production of liquids from shale rock plays such as the Eagle Ford Play in Texas. The sizes of pores in shales holding the oil are believed to be of the order of nano meters. The fluids present in such small nano pores have different properties compared to properties measured in bulk. Fluid saturation pressures at given temperatures – bubble points for oils and dew points for condensates in the nano pores are affected by the influence of pore walls in the vicinity of the fluid molecules. Moreover, kerogen, a cross linked polymer like organic matter present in the shale rocks, absorbs a part of oil present in these rocks. Such interactions with kerogen changes the pressure-volume-temperature (PVT) properties of oil in association with kerogen. Approach to bubble point or dew point influences the proportion of liquid or gas produced from a given well, and thus impacts the economic viability. Hence an accurate measure of saturation pressures is important. The traditional equation of state fails to add the complexity of pore wall interaction in calculating thermodynamics properties of confined fluids. It is important to understand how liquids are produced from ultra-low permeability rocks so that production rates and recovery can be optimized. Apart from the fluid properties, the understanding of transport of fluids through the organic and inorganic matrix in rocks is valuable for economic exploitation of oil and gas from shales.

Current research includes study of thermodynamics of solvent (oil) – kerogen (polymer) interaction, Molecular Dynamics simulations for assessment of hydrocarbon storage in kerogen, Monte Carlo simulations in Gibbs and Grand Canonical ensemble for understanding of flow and phase behavior of confined fluids, experiments conducted on isolated kerogen using Differential Scanning Calorimetry (DSC) to understand swelling of kerogen and associated affect on pressure-volume-properties of oil and fabrication of nano fluidics channels for the study of fluid flow in nano channels of tight rocks. In this paper we describe thermodynamic models, molecular simulations and experiments using DSC and in well characterized fabricated nano channels to understand effect of presence of cross linked polymer and confinement on flow and phase behavior of fluids. The results from Molecular Dynamics Simulations show that the kerogen interacts with hydrocarbons. The subsequent thermodynamic modeling demonstrate that kerogen-oil interaction suppresses the bubble point of oil in shales when compared to the bubble point of oil in the bulk. It was observed that for a type II kerogen with hydrogen index (HI) in range of 450-500 mg/g TOC, a suppression of over 1700 Psia in bubble point pressure (at 400 ^{o}F) was observed for in-situ oil from the original bubble point of 3950 Psia (at 400 ^{o}F) of produced 'free' oil in Eagle Ford Shale oil play.

The possibility of enhanced oil recovery through CO2 sequestration is also studied in this research. The self diffusivities of CO_2 were found to be in order of 10^{-9} m^2/s in kerogen. The research demonstrates that CO_2 has more affinity for adsorption in kerogen like organic macromolecules compared to CH_4. The skill sets developed during this research include but are not limited to thermodynamic modeling, molecular simulations of cross-linked polymers and protein like complex macromolecules with different fluids and fabrication of nano – microfluidics devices. Such skills are easily transferrable to the fields of bio-medical engineering, polymer science, material engineering and semi-conductors.

474453 Highly Active and Durable Extended Surface Electrocatalysts

Abstract:

Highly Active and Durable Extended Surface Electrocatalysts

Shaun M. Alia, Staff Scientist, Chemical and Materials Sciences Center, National Renewable Energy Laboratory

Research Interests:

In the near future, fuel cell vehicles will become cost competitive with other drive train and fuel platforms. The commercial impact of proton exchange membrane fuel cells (PEMFCs), however, is still limited in part by catalyst cost, where the catalyst layer accounts for roughly half of fuel cell costs. Although platinum nanoparticles are commonly used due to high surface area and moderate activity, a reduction in catalyst loading or cost can improve PEMFC prospects and potentially accelerate its progress as a commercial product. In addition to catalyst cost, concerns have arisen due to the durability of the catalyst type.

Extended thin films have been studied as oxygen reduction catalysts since extended platinum surfaces generally offer site-specific activities an order of magnitude greater than nanoparticles. Extended nanostructures also avoid carbon as a support and may offer a long-term durability benefit. The catalyst type, however, has traditionally been limited by low surface area (10 20 m^2 g ^{1}), and improvements are typically accomplished through an alloying effect improving site-specific activity. In this talk, progress in developing PEMFC catalysts through various deposition techniques will be presented. This work includes platinum-nickel nanowires, which have surface areas in excess of 90 m^2 g ^{1} and mass activities an order of magnitude greater than platinum nanoparticles in half-cell testing.

Recently, aspects of fuel cell catalyst development have been applied to hydrogen-producing electrolysis technologies.

Although the cost of electrolyzers is currently dominated by feedstock costs, capital and catalyst costs will become increasingly vital as the power input shifts to renewable, intermittent sources of energy. Advancement in water splitting will be presented, including the development of extended iridium nanostructures in the oxygen evolution reaction. In cases, these materials produce activities an order of magnitude greater than commercial nanoparticles while offering a potentially significant durability advantage.

Teaching Interests:

My research is in electrochemistry and the development of nanomaterials for energy applications and my teaching experience was attained in chemistry and chemical engineering departments. The breadth of my experiences has provided preparation for teaching a variety of courses. While I am capable of teaching any course in the core curriculum, I am particularly interested in teaching kinetics and transport. Teaching undergraduate or graduate electives is also an interest, on areas aligned with my research, including electrochemistry as it relates to catalyst development. The inclusion of courses related to electrochemistry can inspire students to contribute to the advancement of the field and embraces the spirit of chemical engineering, an ever-evolving discipline at the forefront of technological advancement.

In previous experience as a teaching assistant, I emphasized student participation by having them periodically instruct each other. By running discussion sections and being a guest lecturer, experience was also attained in the practical aspects of teaching, including preparing lectures, lesson plans, and exams. As a faculty candidate, I am interested in promoting student involvement and providing practical applications to support concepts in the form of video or demonstrations. In addition to teaching core chemical engineering courses, I also am interested in bringing energy and nanomaterial related courses into the curriculum.

475433 Electrokinetic Transport in Porous Media for Energy and Environmental Applications

Abstract:

Research Interests: Electrokinetics, Transport Phenomena, Computational Science and Engineering, High Performance Computing, Random Materials

Teaching Interests: Fluid Mechanics, Thermodynamics, Numerical Analysis, Mathematical Physics

Understanding the transport phenomena and electrochemical processes in porous media is of great importance in many applications. Examples include energy applications such as in batteries and supercapacitors, electrochemical processes such as electro-catalysis, as well as industrial scale water desalination or geological scale fluid flow problems in underground reservoirs such as in oil recovery. Although these examples span several orders of magnitude in both length and time scales, the fundamental picture of electrokinetic phenomena is remarkably relevant in almost all of them. Here, we present some recent works on two aspects of electrokinetic transport in porous media.

First, we discuss the surface conduction phenomena and its effect on shortening the charging kinetics in porous electrodes. In most models of electrokinetic transport in porous media, the role of pore microstructure is often neglected and the transport is descried via homogenized, spatially averaged quantities. Inspired by detailed Direct Numerical Simulation (DNS) of the Poisson-Nernst-Planck (PNP) equations at the pore-scale, we show that the pore microstructure can have significant effect on the charging kinetics. Specifically, we observe that in the presence of large applied potentials, the surface conductivity of the Electric Double Layer (EDL) becomes considerably large, competing with that of the bulk of electrolyte. This allows for passage of large surface current through the EDL, which effectively acts to â€œshort-circuitâ€ the bulk and significantly enhance the charging kinetics. We believe this discovery could be utilized to engineer better preforming electrodes for high power density applications.

Second, we describe the effects of electric field on the viscous fingering instability in Hele-Shaw cells. Large scale flow problems in porous media, such as those encountered in underground oil reservoirs, are typically described via Darcyâ€™s law. However, it is well known that many underground rock formations contain surface groups and minerals that dissociate in the presence of water. Convection of these charges by the pressure driven flow can then set up streaming current and streaming potential that affects the flow. Furthermore, electric fields are sometimes used to enhance the oil recovery, e.g. by reducing the oil viscosity through electro-thermal heating or enhancing flow rate via electro-osmotic flow. Thus, a full description of fluid flow in the presence of electric field must also include the electrokinetic phenomena.

It is well known that the moving interface between two fluids in a porous medium becomes unstable if pushed by the less viscous fluid. Here, we report on the role of electrokinetic phenomena on stability of these viscous fronts in Hele-Shaw cells using analytic as well as numerical approaches. Interestingly, we find that instability could be suppressed if the right physical conditions are met or otherwise enhanced, leading to greater mixing of two fluids. These findings might be of interest to researchers working in the area of Electrically Enhanced Oil Recovery.

475796 Modeling Multiphysics Transport Phenomena and Device Design for Solar-Fuel Technologies

Abstract:

2ndyear Postdoctoral Fellow

Research Motivation and Interests:

Imagine if we could take a ***leaf*** from nature's book to convert and store available sunlight, water and carbon dioxide in stable chemical bonds of hydrogen and hydrocarbon compounds. These solar-fuels could provide carbon-neutral alternatives to the fossil fuel dominated energy sector with enormous potential for large-scale storage and long-term transportation. Ingenious fundamental science and device-scale engineering advancements are needed to efficiently harness, convert and store solar energy along with favorable economic drivers to facilitate global technological adoption. My research interests and strengths are to develop mathematical models and high fidelity computational simulations closely connected with experimental investigations to design scalable, stable and cost-effective integrated solar energy conversion and storage devices. Specifically, I am interested in probing the complex interactions of the various physical phenomena – optical absorption, heat and mass transfer, heterogeneous chemical reactions and interfacial electron transport to optimize the materials-to-device scale performance of high- and low- temperature solar fuel generators.

Awards: Doctoral Dissertation Fellowship awarded by the Graduate School at the University of Minnesota for 175 Ph.D. candidates from a campus wide competition.

Postdoctoral Project: ***"Design of a Particle-Suspension Reactor for Photoelectrochemical Solar Water Splitting"***

Supervised by Dr. Adam Z. Weber, Lawrence Berkeley National Laboratory, and Prof. Shane Ardo, University of California, Irvine.

PhD Dissertation: ***"Transport and Chemical Phenomena in a Solar Reactor to Split Water and Carbon Dioxide to Produce Synthesis Gas"***

Supervised by Prof. Jane H. Davidson, Christenson Chair in Renewable Energy, University of Minnesota, Twin Cities

Research Interests:

My graduate and postdoctoral research work has helped me acquire an interdisciplinary background in the areas of thermal sciences and fluid mechanics, semiconductor device physics and electrochemistry. I have gained substantial expertise in developing numerical transport models using in-house codes as well as commercial softwares to simulate interdependent transport processes for device-scale performance optimization. I am adept at parallelizing computational codes to implement them on high performance computers (HPCs), which is especially beneficial for three-dimensional and transient problems. Beyond continuum modeling approaches, I am also experienced with Monte Carlo simulations for radiative transport and meso-scale Lattice-Boltzmann approaches to simulate sedimentation of bacterial swimmers. As part of the reactor design team at the University of Minnesota, we designed and tested a 4 kW, high-temperature solar reactor integrated with gas phase heat recovery to operate at 1500 °C. The reactor implements an isothermal ceria redox cycle to split water and carbon dioxide to produce synthesis gas. The prototype development and reactor operation has given me an opportunity to view a heat and mass transfer, reaction-engineering problem from the eyes of materials scientists and structural mechanics researchers. In my current research, I am trying to understand the intricacies involved in semiconductor physics and interfacial electrochemical reactions in particulate suspensions. All my research projects so far have had a significant experimental and prototyping component, which has taught me the value of exploiting combined numerical-experimental approaches to connect the dots in multi-physics problems with varying length and time scales. This experience has also provided me with a versatile knowledge of experimental toolkits – thermocouple instrumentation and control, gas analysis equipment (RLGA and mass spectrometer), UV-VIS spectrophotometer, dynamic light scattering and cyclic voltammetry, to name a few.

Teaching Interests:

I have been a Teaching Assistant for over three years of my graduate study in undergraduate physics (statics and mechanics, electricity and magnetism) and mechanical engineering core courses (fluid mechanics, heat transfer, thermodynamics and measurements laboratory). I have particularly benefited from my experiences of running laboratory classes that involved designing experimental test stations and control devices for teaching fundamental transport phenomena concepts. I've also had the opportunity to mentor graduate students in my research lab to train them on performing numerical simulations and have worked closely with them to write and orally present scientific results. Outside of professional work, I strive to encourage younger students' participation in science and mathematics by volunteering in high schools to help students with basic algebra, geometry and physical sciences.

Future Direction:

As faculty I would like to apply my expertise in transport modeling for designing solar-fuel generators and other energy conversion systems to further the fundamental understanding of the interactions between various physical phenomena, and use this knowledge to offer device-scale engineering solutions. I hope to integrate my experiences from my PhD and postdoctoral work to more closely investigate full-spectrum utilization and heat management in integrated photoelectrochemical devices subjected to concentrated solar radiation. Concentrated solar power provides the benefit of higher power input per unit area of the device but also causes excess heating that could deteriorate the performance of the semiconductor materials. However, intelligent material selection and device design could potentially redirect the ultraviolet and visible portions of the solar spectrum towards water and carbon-dioxide splitting and channel the high-quality heat in the infrared wavelengths for heat storage.

Besides renewable solar fuel applications, I am also interested in solar gasification of biomass and other-waste materials, redox flow batteries, water treatment and pollution abatement with photocatalysts, thermal energy storage in phase change materials and modeling heat and air flow to optimize building energy requirements. I envision that developing numerical models of varying complexity and length scales along with carefully designed experiments can result in optimized performances, irrespective of the end application.

Selected Publications

[1] **Bala Chandran R.**, and Davidson J. H., 2016, "Model of transport and chemical kinetics in a solar thermochemical reactor to split carbon dioxide," Chem. Eng. Sci., **146**, pp. 302–315.

[2] **Bala Chandran R.,** De Smith R. M., and Davidson J. H., 2015, "Model of an integrated solar thermochemical reactor/reticulated ceramic foam heat exchanger for gas-phase heat recovery," Int. J. Heat Mass Transf., **81**, pp. 404–414.

[3] Hathaway B. J., **Bala Chandran R.**, Sedler S., Thomas D., Gladen A., Chase T., and Davidson J. H., 2015, "Effect of Flow Rates on Operation of a Solar Thermochemical Reactor for Splitting CO_2 Via the Isothermal Ceria Redox Cycle," J. Sol. Energy Eng., **138**(1), p. 011007.

[4] **Bala Chandran R.,** Bader R., and Lipiński W., 2015, "Transient heat and mass transfer analysis in a porous ceria structure of a novel solar redox reactor," Int. J. Therm. Sci., **92**, pp. 138–149.

[5] Bader R., **Bala Chandran R.**, Venstrom L. J., Sedler S. J., Krenzke P. T., De Smith R. M., Banerjee A., Chase T. R., Davidson J. H., and Lipinski W., 2015, "Design of a Solar Reactor to Split CO_2 Via Isothermal Redox Cycling of Ceria," J. Sol. Energy Eng., **137**(3), p. 031007.

476218 Understanding and Optimizing Energy and Mass Transport in Porous Materials for Water, Energy, and Thermal Management Applications

Abstract:

Research Interests:

Porous materials are ubiquitous and vital in a wide range of applications. Transport of energy, mass, and charge to the large fluid/solid interfacial area provided by these materials often controls the behavior of such systems. My research examines these transport phenomena as well as the functionalization of the surfaces themselves to optimize performance in several applications including toxin removal from water, energy storage in electrochemical capacitors, and dissipation of extreme heat fluxes from next generation electronic systems.

Part of my ongoing research addresses the challenge of water purification via capacitive deionization (CDI), an emerging technology which removes ionic contaminants from water via electrostatic interactions with the porous electrodes of a capacitor. Specifically, we focus on inverted CDI approaches using surface functionalization of porous electrodes to produce enhanced affinity for a variety of toxins. The porous electrodes act as adsorbents with selectivity for certain toxins. These adsorbents may then be regenerated by the action of an applied electric potential, which drives off the charged toxin species (as compared to traditional CDI where ions are attracted to the electrode surface by an applied potential). We have shown the ability to remove common contaminants from impaired input water streams to levels well below safe drinking water limits and to later expel these contaminants from the porous media without the need for the attendant waste disposal and cost concerns of chemical regeneration. My future research aims toward developing novel materials for better selectivity of difficult to treat impurities and the design of cells to maximize throughput and efficiency.

Another application closely related to CDI is energy storage in electrochemical capacitors. Recent advances in electrode materials provide great potential to enhance the energy density of these systems based on increases in material capacitance and pseudo-capacitance. However, due to the resulting potential for electrolyte depletion based on intrinsic solubility limits or desired operational regimes, the response of these systems involves a complex interplay of electromigration and diffusive transport of electrolyte species throughout the porous structures of the electrodes and spacer. We have developed parametric transport models including electrolyte depletion, capable of accounting for variations in cell geometry, electrolyte composition, and electrode material capacitance, resistivity, and structure. I will extend these models to provide guidance for cell design, electrode material choice, and electrolyte composition leading to electrochemical capacitors capable of addressing arising demands in grid energy storage and electric vehicle applications.

Beyond their use in electrochemical systems, porous materials have great potential in thermal management systems, such as heat pipes. We have studied energy and mass transport in two-phase cooling systems allowing dissipation of extreme heat fluxes, which is key to enabling dramatic performance improvements in a wide variety of electronic components including microwave amplifiers, power transistors in inverters for renewable power generation and electric vehicle applications, and 3D integrated microelectronics. We have produced thin conformal porous copper films with precisely defined microstructure using templated electrodeposition techniques. These structures with high surface area and thermal conductivity support heat dissipation via boiling of fluxes in excess of 1200 W/cm^2 at superheats less than 10 K from hot spots in electronic devices. We have also integrated these films with extended surfaces formed from laser micromachined diamond to create heat sinks capable of extending this performance to large areas. In future, I will focus on improving the mechanistic understanding of two phase mass and energy transport during boiling in porous media and using this knowledge to design two-phase cooling systems employing porous media to solve current and emerging thermal management challenges.

Teaching Interests:

I am prepared to teach a variety of courses in the chemical and mechanical engineering and materials science curriculums. I would be excited to teach materials related courses as well as thermodynamics, heat and mass transfer, and fluid dynamics. I am also interested in developing a new graduate course or seminar providing an integrated treatment of fluid, thermal, charge, and chemical transport in porous media along with relevant surface science.

478649 Nanoscale Materials for Energy Storage and Conversion

Abstract:

Research Interests:

High efficiency energy storage and conversion are among the main challenges facing governments and industries that supply the energy needs of modern societies. Progress in the development and manufacturing of new low-dimensional and nanostructured materials, which can efficiently store and convert energy, may help us overcome these challenges – if we can design them correctly. In this regard, looking into the molecular structure and dynamics of new materials and characterizing their properties for such applications is an advantage of atomistic modeling.

My research goal is to adopt and utilize molecular modeling-based methods to characterize and predict the properties/behavior of novel materials/systems, and ultimately improve the efficiency of energy storage and conversion systems. I propose to create a multidisciplinary research program focusing on characterizing and predicting thermo-physical, transport and thermo-electrical properties of newly developed materials for energy applications.

Postdoctoral Project: "*Thermal transport in Metal Organic Frameworks (MOFs)*"

Under supervision of Professor C. Wilmer and Professor A. McGaughey

Visiting Scholar: "*Thermoelectric properties of single layer* MoS_2 *by using ab initio calculations*"

Under supervision of Professor S. Sinha

PhD Dissertation: "*Molecular-Level Modeling of Thermal Transport Mechanisms within Carbon Nanotube/Graphene-based Nanostructure-Enhanced Phase Change Materials*"

Under supervision of Professor J. M. Khodadadi and Professor P. Keblinski, RPI

MS Dissertation: "*Simulation of Turbulent Vortex Shedding Past a Square Cylinder near a Wall*"

Under supervision of Professor M. Raisee

BS Dissertation: "*Analysis of Turbulent Flow through a Channel using 2nd and 3rd Grade Fluids*"

Under supervision of Professor G. Atefi

Past and Current Research:

My academic career path has been a blend of several fields of science and engineering. My formal training was in simulation, which I have learned on many different physical scales: Density Functional Theory (femto), Molecular Dynamics (nano), and Computational Fluid Dynamics (continuum).

My PhD research project was devoted to non-continuum-based modeling of the effects of adding nanoparticles to fluids/phase change materials (PCM) and the heat transport within the resulting nanoparticle suspensions/composites. I adopted the methodology within molecular dynamics simulation approach to calculate the thermal conductivity of multi-component systems. Using the established method and other advanced molecular dynamics-based methods, I studied heat transfer in nanofluids and nanostructure-enhanced PCM (NePCM). I found that among different reported mechanisms of heat transfer in nanofluids, only clustering of nanoparticles can result in high enhancement in thermal transport. For PCM (here, paraffin), introducing CNT and graphene sheets promotes aligning of the molecules in the direction parallel to the CNT axis and graphene surfaces, respectively and consequently leads to considerable enhancements in its thermal conductivity along those directions. Moreover, in collaboration with Dr. Jackson's group at Auburn University, by using molecular modeling, I found that presence of nanoparticles enhances the thin film elasto-hydrodynamic lubrication.

During my one-year stay at UIUC, by utilizing first principle and phonon calculations, I investigated the thermoelectric properties of single-layer molybdenum disulfide (SL-MoS_2) (a 2-D material) and found an anomalously large thermoelectric power factor. Moreover, I developed a new methodology in finding phonon relaxation times.

During the past two years at Pitt and CMU, I have been studying the mechanisms of heat transfer in Metal-organic frameworks (MOFs) during gas adsorption using molecular dynamics (MD) and grand canonical Monte Carlo (GCMC) simulations and spectral energy density (SED) calculations. All these research projects have led to important findings so far and have opened up room for new ideas to the nanoscale heat transfer community which call for more research work, some of which will be outlined in the following section.

Furthermore, over these years I have gained deep knowledge of various aspects of different modeling techniques including MD simulation, Density Functional Theory (DFT), Density Functional Perturbation Theory (DFPT), MC, SED, lattice dynamics, thermal conductivity calculations and quantum modeling that are suitable to be used in studying other interdisciplinary nanoscale research problems as well.

Future research plans

Project 1. Low-dimensional thermoelectric materials

Thermoelectric materials provide a direct way of converting waste heat and solar energy into electricity. For instance, they can be used to convert waste heat in vehicles into electricity. However, lack of a scalable source of thermoelectric materials having the required conversion efficiency has prevented commercializing these devices. One way to overcome the problem of low efficiency is to use low-dimensional materials. Recent developments in nanotechnology have made it possible to adjust the electrical and thermal properties of materials in such a way that they provide a higher amount of thermoelectric figure of merit.

I will devote one part of my research to study thermoelectric properties of 1- and 2-D materials (e.g., nanotubes, nanowires, single and few-layer materials) whose electronic structure and transport properties potentially allow designing more efficient thermoelectric systems. In this general research project, I will focus my research on three properties of materials required for evaluating their thermoelectric efficiency: (1) Seebeck coefficient, (2) electrical conductivity and (3) thermal conductivity. For evaluating the first two properties one needs information on electron relaxation times. I will use ab initio calculations to obtain the relaxation times for phonons and electrons, which are the parameters needed to evaluate the transport properties involved in characterizing the efficiency of thermoelectric materials.

Project 2. Thermal properties of metal-organic frameworks

Metal-organic frameworks (MOFs), porous materials with extremely high internal surface areas, can adsorb large amount of gases and can be used for natural gas storage (e.g. in passenger vehicles). However, one challenge with MOFs is the heat generated during gas adsorption which leads to high temperatures, and in turn, prevents gas adsorption and reduces storage capacity. In principle, designing systems that could dissipate heat more quickly can allow for faster filling times. However, designing porous materials with better thermal properties requires a detailed understanding of the atomic-scale heat transfer mechanisms, which is still lacking in the literature. I propose a research project aiming to further our understanding of heat transfer in these systems. In this project I will use the adopted state-of-the-art molecular modeling approaches including MD, GCMC and SED calculations to study these systems and predict the needed thermal properties.

Project 3. Thermal transport in nanostructure-enhanced energy storage systems

One of the most convenient ways to store thermal energy is using phase change materials (PCM). PCM offer their sizeable latent heat for storing thermal energy at a constant temperature. However, a weakness of such materials is their relatively low thermal conductivity, which strongly suppresses the energy charge/discharge rates. A challenging option is to suspend highly-conductive particles into PCM, which produces "free-form" mixtures/composites5. This challenging problem is one of my research interests on which I will assign one part of my research group to work. I will use non-equilibrium MD method to characterize the required thermal properties of PCM composites. MD is an excellent candidate for handling the length and time scales that are expected in this problem.

Teaching Interests:

For my future academic career, I am prepared to teach general undergraduate and advanced graduate level courses in chemical Engineering, mechanical engineering and materials science, such as heat transfer, fluid mechanics and thermodynamics. Moreover, I would also like to teach interdisciplinary courses such as "statistical physics" and "transport phenomena".

I believe that with the recent growing demands for smaller scale technologies (e.g. nanotechnology), leading engineering departments need to develop courses related to these areas. During the past years dealing with projects in nanoscience, I have always been thinking of developing a course or courses that can effectively address both basic physics and techniques required in the burgeoning areas for engineering students. I think "nanoscale transport phenomena" and "molecular modeling" could be two of such new proper courses within the department of mechanical engineering.

Selected Publications:

H. Babaei, A. J. H. McGaughey and C. E. Wilmer, *Effect of pore size and shape on the thermal conductivity of metal-organic frameworks,* Chemical Science (2016). DOI: 10.1039/c6sc03704f.

H. Babaei and C. E. Wilmer, *Mechanisms of Heat Transfer in Porous Crystals Containing Adsorbed Gases: Applications to Metal-Organic Frameworks,* Physical Review Letters **116**, 025902 (2016). **PRL Editors' Suggestion**.

H. Babaei, J. M. Khodadadi, and S. Sinha, *Large theoretical thermoelectric power factor of suspended single-layer* MoS_2, Applied Physics Letter **105**, 193901 (2014).

H. Ghaednia, **H. Babaei**, R. L. Jackson, M. J. Bozack, J. M. Khodadadi, *The effect of nanoparticles on thin film elasto-hydrodynamic lubrication*, Applied Physics Letters, **103**(26), 263111 (2013). ***The first two authors contributed equally to this work***.

H. Babaei, P. Keblinski, and J. M. Khodadadi, *Improvement in thermal conductivity of paraffin by adding high aspect-ratio carbon-based nano-fillers,* Physics Letters A, **377**, 1358–1361 (2013).

H. Babaei, P. Keblinski, and J. M. Khodadadi, *A Proof for Insignificant Effect of Brownian Motion-Induced Micro-Convection on Thermal Conductivity of Nanofluids by Utilizing Molecular Dynamics Simulations,* J. Appl. Phys, **113**, 084302 (2013).

J. M. Khodadadi, Liwu Fan and **Hasan Babaei,** *Thermal Conductivity Enhancement of Nanostructure-Based Colloidal Suspensions Utilized as Phase Change Materials for Thermal Energy Storage: A Review,* Renewable & Sustainable Energy Reviews, **24**, 418 (2013).

H. Babaei, P. Keblinski, and J. M. Khodadadi, *Thermal conductivity enhancement of paraffin by increasing the alignment of molecules through adding CNT/Graphene,* International Journal of Heat and Mass Transfer, **58**, 209 (2013).

H. Babaei, P. Keblinski, and J. M. Khodadadi, *Equilibrium Molecular Dynamics Determination of Thermal Conductivity for Multi-Component Systems,* J. Appl. Phys. **112**, 054310 (2012).

M. Raisee, A. Jafari, **H. Babaei**, H. Iacovides, *Two-Dimensional Prediction of Time-Dependent, Turbulent Flow around a Square Cylinder Confined in a Channel*, International Journal for Numerical Methods in Fluids, **62**, 1232 (2010).

Successful Proposals: ACS Petroleum Research Fund, 2015-2018; AL-EPSCoR GRSP Fellowship 2010-2014 ($75,000)

476141 Harnessing Micro-Scale Transport Processes for Advanced Healthcare, Personalized Medicine and Point of Care Diagnostics

Abstract:

Research Interests:

Recent outbreaks like Zika and Ebola highlights the challenges associated with pathogen diagnostics in the developing world. With the outbreak in Africa, and isolated cases on other continents, the need for an affordable, rapid and portable diagnostic solution has been repeatedly stressed and is one of the most critical issues confronting global health. Unfortunately, the current conventional PCR instrumentation needed to perform "gold standard" DNA-based diagnostic tests is bulky, slow, and expensive, making it unsuitable for resource limited settings in developing countries where dedicated laboratory facilities are not available. Advances in micro-fluidics and smartphone based technology has paved way for novel implementations of traditional molecular diagnostic techniques.

Recently, it has been shown that convective flow fields can be harnessed to perform PCR. My research led me to investigate the underlying physics of these thermally driven micro-scale Rayleigh-Bénard convective flows. This new understanding revealed how a subset of these flow fields exhibit chaotic advection that greatly accelerate temperature driven biochemical reactions such as PCR [1-4]. Having found the optimal design parameters for convective PCR, I coupled this novel isothermal setup with a versatile smartphone based detection unit and an integrated image analysis app [5-7]. This greatly simplifies the design, enabling us to integrate such biochemical analysis platforms with consumer-class quadcopter drones for rapid deployment of nucleic acid based diagnostics[8]. The ability to perform rapid in-flight assays with smartphone connectivity eliminates delays between sample collection and analysis so that test results can be delivered in minutes, suggesting new possibilities for drone-based systems to function in broader and more sophisticated roles beyond cargo transport and imaging.

The surprising interplay between reactions and micro-scale convective flows led me to consider adaptations beyond PCR. Specifically, we demonstrate that such flows, naturally established over a broad range of hydrothermally relevant pore sizes, function as highly efficient conveyors to continually shuttle molecular precursors from the bulk fluid to targeted locations on the solid boundaries, enabling greatly accelerated chemical synthesis. Insights from this study has the potential to provide a breakthrough in our understanding of the fundamental biochemical processes underlying the origin of life [9, 10]. Another aspect of my research involves the hydrodynamical interaction between micron sized particles in intricate flow fields. My computational fluid dynamics models have thrown light on how the micrluidic flow fields can be tune fluid-particle interactions[9, 11].

At Sandia National laboratories, I have continued the development novel molecular diagnotics techniques for point of care applications, which involves an innovative approach to pathogen detection via reverse transcription loop-mediated isothermal amplification (RT-LAMP). A new kinetic model for LAMP reactions is formulated in order to shine light on the underlying reaction pathways. The new insights gained from the model are then used for efficient primer design and determining optimal reaction parameters. The LAMP technique is then coupled with an extremely simple and user-friendly smartphone enabled detection platform capable of detecting bright, unambiguous fluorescence signals produced. Finally, I have made significant contribution to advance the development of "Smart traps" – a next generation of rugged yet inexpensive robots to perform autonomous field surveillance of arboviruses. These traps are able to analyze saliva from vector-borne mosquitoes with daily LAMP assays and transmit data between each other over a wireless XBEE network. The ability to deliver performance comparable to that of current generation systems while simultaneously providing an order of magnitude reduction in cost and turnaround time has the potential to greatly expand the use of nucleic acid based detection assays by moving them out of the laboratory and into settings where they are needed most.

Teaching Interests:

One of the most fascinating subjects I have ever come across would have to be fluid mechanics and transport processes as it is conceptually quite rich and at the same time hosts a tremendous opportunity to incorporate hands on learning experience. There is a need for innovative educational experiences that unify and reinforce fundamental principles at the interface between the physical, chemical, and life sciences. These experiences empower and excite students by helping them recognize how interdisciplinary knowledge can be applied to develop new products and technologies that benefit society. Microfluidics offers an incredibly versatile tool to address this need. In one of my educational paper[1, 12], we describe our efforts to create innovative hands-on activities that introduce chemical engineering students to molecular biology by challenging them to harness microscale natural convection phenomena to perform DNA replication via the polymerase chain reaction (PCR). Experimentally, we have constructed convective PCR stations incorporating a simple design for loading and mounting cylindrical microfluidic reactors between independently controlled thermal plates. A portable motion analysis microscope enables flow patterns inside the convective reactors to be directly visualized using fluorescent bead tracers. We have also developed a hands-on computational fluid dynamics (CFD) exercise based on modeling microscale thermal convection to identify optimal geometries for DNA replication. A cognitive assessment reveals that these activities strongly impact student learning in a positive way.

1. Priye, A., Y.A. Hassan, and V.M. Ugaz, *Education: DNA replication using microscale natural convection.* Lab on a chip, 2012.

2. Priye, A., Y.A. Hassan, and V.M. Ugaz, *Microscale chaotic advection enables robust convective DNA replication.* Analytical chemistry, 2013. **85**(21): p. 10536-10541.

3. Priye, A., et al., *Selecting 3D Chaotic flow states for Accelerated DNA replication in micro-scale*

convective PCR. 15th International Conference on Miniaturized Systems for Chemistry and Life Sciences, 2011. **15**.

4. Priye, A. and V.M. Ugaz, *Convective PCR Thermocycling with Smartphone-Based Detection: A Versatile Platform for Rapid, Inexpensive, and Robust Mobile Diagnostics*, in *Microfluidic Methods for Molecular Biology*. 2016, Springer International Publishing. p. 55-69.

5. Jensen, M., et al., *ANALYTICAL INSTRUMENT SYSTEMS*. 2016, US Patent 20,160,025,630.

6. Priye, A. and V. Ugaz, *PCR to Go*. 2014, Apple iOS appstore https://itunes.apple.com/us/app/pcr-to-go/id909227041?mt=8: Apple iOS appstore.

7. Priye, A. and V. Ugaz, *DNA-to-go: A portable smartphone-enabled PCR assay platform.* arXiv preprint arXiv:1606.02252, 2016.

8. Priye, A., et al., *Lab-on-a-drone: toward pinpoint deployment of smartphone-enabled nucleic acid-based diagnostics for mobile health care.* Analytical chemistry, 2016. **88**(9): p. 4651-4660.

9. Priye, A., *Micro-Scale Complex Flows Enables Robust DNA Replication, Enhanced Transport and Tunable Fluid-Particle Interactions.* 2015.

10. Priye, A., Y.A. Hassan, and V.M. Ugaz, *Thermally-targeted adsorption and enrichment in microscale hydrothermal pore environments.* 17th International Conference on Miniaturized Systems for Chemistry and Life Sciences, 2013. **17**.

11. Priye, A. and W.H. Marlow, *Computations of Lifshitz–van der Waals interaction energies between irregular particles and surfaces at all separations for resuspension modelling.* Journal of Physics D: Applied Physics, 2013. **46**(42): p. 425306.

12. Ugaz, V.M., A. Priye, and Y.A. Hassan, *DNA to Go: A Do-it-Yourself PCR Thermocycler Lab.* American Society for Engineering Education, 2012.

476145 Leveraging Systems-Based Assessment to Understand Research and Development Needs for Emerging Energy Processes

Abstract:

The evaluation of an emerging energy process is often driven by the development of models which utilize process science and engineering fundamentals. The objective of process models is to facilitate an understanding of research and development opportunities, and to prepare a framework for analyzing experimental information. The evaluation scenarios are often characterized by limited data availability owing to their exploratory nature, and leveraging synergies from analogous processes is often valuable in formulating the first steps. In research and development organizations these activities are often categorized as research guidance, process evaluation and planning, or as exploratory research.

Research Interests:

Conceptual process design in combination with techno-economic and infrastructure integration analysis is a robust methodology which helps in understanding critical initiatives for fostering research and development. At the National Renewable Energy Laboratory's National Bioenergy Center, I have worked on the development of process models catering to various bioenergy processes of interest enabled by funding from the U.S. Department of Energy's Bioenergy Technologies Office. These include co-conversion of natural gas and biomass to fuels, catalytic fast pyrolysis of biomass, co-processing of petroleum refinery intermediates and bio-based feedstocks in hydroprocessing, and blending of petroleum and bio-based intermediates. These efforts are significantly important as the bioenergy industry is focusing on the near-term development of conversion pathways from biomass feedstocks to infrastructure-compatible fuel blendstocks. One of the inspirations to contribute towards on-going infrastructure integration analysis efforts emerged from my industrial research and development experience in India (Bharat Petroleum Corporation Limited).

My doctoral research at the University of Utah was focused on the development of conceptual process design principles for a 200 kWth process development unit for Chemical-looping with Uncoupling (CLOU). CLOU is a high temperature gas-solid fluidized bed process for solid fuel combustion to facilitate carbon capture utilization and sequestration (CCUS). The results of my research served as one of the foundational steps in facilitating an understanding of CLOU process fundamentals. The research was funded through the U.S. Department of Energy's Office of Fossil Energy (under Award No. DE-NT0005015).

Teaching Interests:

In addition to my interests in research in energy and sustainable transportation (bioenergy, petroleum and solid fuels), I am highly interested in chemical engineering education. During my post-doctoral research at NREL, I have mentored students, and was awarded an Outstanding Mentor Award by the U.S. Department of Energy, Office of Science's Science Undergraduate Laboratory Internships (SULI) program in 2014. During my doctoral degree, I had served as a Teaching Assistant Scholar towards developing a framework for undergraduate students titled "Inspirations for Innovation" – (http://ctle.utah.edu/tas/projects/asad-sahir.php) with a perspective on product and process design, and have mentored undergraduate students for research. I have also organized workshops at the University of Utah namely "Exploring the Human Dimension as a Teaching Assistant (TA) in the STEM Classroom" in 2013 and "International Graduate Learning, Teaching and Research (IGLTR) Workshop" in 2012.

This poster is a showcase of my research and academic interests which are driven by formulating an understanding the initial knowledge deficiencies associated with the design of an emerging energy process, and subsequently developing bridges through process modeling and chemical engineering principles. As a prospective faculty candidate, I intend to leverage my unique research experience at a national laboratory, industry and during my doctoral degree to help students appreciate core chemical engineering principles, and subsequently inspire and enable them to chart out a course in energy innovation for their community.

468388 Control of Fe-Ni Nanoparticle Activity for the Oxygen Evolution Reaction (OER) and Methanol Electrooxidation

Abstract:

The electrolysis of water for hydrogen generation and the electrochemical oxidation of alternative fuels such as methanol for alkaline fuel cells currently relies on the use of precious metal catalysts. However, the scarcity and high cost of these precious metals lead to the search for earth-abundant alternatives with similar electrocatalytic properties. Key challenges in the development of catalysts for the oxygen evolution half reaction (OER) of water electrolysis remain; these challenges include reducing the overpotential of the reaction and developing non-precious metal catalysts that are stable and active with minimal mass transport limitations. For direct electrooxidation of alternative fuels such as methanol, non-precious metal alternatives to precious metals such as platinum and palladium are often much less active. Among the non-precious metals, multimetallic nickel-based catalysts show particular promise for their high stability and high activity for the oxygen evolution reaction (OER) half-reaction of water electrolysis. In particular, it has recently been demonstrated by Trotochaud, Burke, Boettcher, and co-authors that iron incorporation into nickel hydroxide and cobalt hydroxide catalysts causes a significant increase in the OER activity of the catalyst [1-3]. It is now understood that the iron is in fact the active site of such bimetallic catalysts [2]. Additionally, nickel has exhibited high activity for methanol oxidation, particularly when in the form of a bimetallic catalyst such as NiPd [4]. However, promising active catalysts that are comprised entirely of non-precious metals such as nickel and iron have not been often reported for alternative fuel electrooxidation. Furthermore, there is a need for the development of nanostructured catalysts that can address mass transport limitations inherent in bulk or film-based catalyst materials [2]. As such, there continues to be an opportunity in alkaline electrocatalyst development to discover and understand nanostructured catalyst materials for both OER and alternative fuel electrooxidation.

In this presentation, our results on the development and optimization of an FeNi hydroxide core-shell nanoparticle catalyst will be discussed for both OER and methanol electrooxidation. Through control of critical synthesis parameters, such as the addition rate of reducing agent, ratio of reducing agent to iron, and reaction time for nanoparticle synthesis, we have demonstrated the ability to tune nanoparticle catalyst activity for either OER or methanol electrooxidation. Results for maximum OER current as a function of the rate of addition of reducing agent sodium borohydride illustrate the sensitivity of catalyst performance to changes in synthesis parameters. Characterization of nanoparticle structure and phase will be discussed and correlated to electrochemical performance. Results suggest that tuning of the crystalline structure and order are critical to electrocatalyst performance optimization, as well as the presence of oxide and hydroxide.

References

[1] L. Trotochaud, S.L. Young, J.K. Ranney, S.W. Boettcher, Nickel-iron oxyhydroxide oxygen-evolution electrocatalysts: The role of intentional and incidental iron incorporation, J. Am. Chem. Soc., 136 (2014) 6744-6753.

[2] M.S. Burke, L.J. Enman, A.S. Batchellor, S.H. Zou, S.W. Boettcher, Oxygen evolution reaction electrocatalysis on transition metal oxides and (oxy)hydroxides: Activity trends and design principles, Chemistry of Materials, 27 (2015) 7549-7558.

[3] M.S. Burke, M.G. Kast, L. Trotochaud, A.M. Smith, S.W. Boettcher, Cobalt-iron (oxy)hydroxide oxygen evolution electrocatalysts: The role of structure and composition on activity, stability, and mechanism, J. Am. Chem. Soc., 137 (2015) 3638-3648.

[4] A. Dutta, J. Datta, Energy efficient role of Ni/NiO in PdNi nano catalyst used in alkaline DEFC, J. Mater. Chem. A, 2 (2014) 3237-3250.

475202 Transition Metal-Nitrogen-Carbon Electrocatalysts for Oxygen Reduction Reaction

Abstract:

UNM has developed the Sacrificial Support Method (SSM) as a main approach for templated synthesis of hierarchically structured electrocatalysts materials. In this method the catalysts precursors are being absorbed on, impregnated within or mechanically mixed with the support (usually mono-dispersed or meso-structured structured silica), thermally processed (pyrolyzed) and then the silica support is removed by etching (in KOH or HF) to live the open frame structure of a "self-supported" material that consists of the catalysts only. Such hierarchical structures are advantageous in enhancement of the fuel cell performance since they correspond to the different levels of transport in the corrugated electrode matrixes. A wide variety of materials can be made by these methods in which not only the composition but also the microstructure can be varied. It is the combination of these attributes - control over microstructure at a number of different length scales and composition, simultaneously - that is extremely important to the performance of the electrocatalyst materials in a fuel cells.

The makeup and structure of the active site/sites of the Platinum Group Metal-free (PGM-free) ORR electro-catalysts, including geometry (coordination) and chemistry (composition and oxidation state) remain contentious even after 50 years of research. There is an emerging agreement that iron and nitrogen functionalities, displayed on the surface if the carbonaceous substrate/support, govern ORR activity. This is often combined with a broadly accepted hypothesis that micro-porous surface area plays a critical role forming the active site. Candidate structures participating in ORR include multitudes of nitrogen defect motifs in the carbon matrix of different degrees of graphitization, with metal incorporated as metal nano-particles, corresponding (native) oxides and/or as atomically dispersed, oxidized metal species, linked (coordinated) to nitrogen defects in carbonaceous matrix in a variety of possible configurations. These consist of in-plane defects such as graphitic nitrogen and Fe-coordinated to three or four nitrogen atoms and a plurality of possible edge sites such as pyridinic, pyrrolic, quaternary and $Fe-N_2/Fe-N_x$ sites. In this study, we have analyzed more than 45 distinct M-N-C electrocatalysts synthesized from three different families of precursors, such as (i) N-containing polymer, (ii) N-chelating macrocycles and (iii) N-containing small organic molecules (amines). Catalysts were evaluated and analyzed structurally using exactly the same protocol for deriving structure-to-property relationships. We have identified possible active sites participating in different ORR pathways: (1) metal-free electrocatalysts support partial reduction of O_2 to H_2O_2; (2) pyrrolic nitrogen acts as a site for partial O_2 reduction to H_2O_2; (3) pyridinic nitrogen displays catalytic activity in reducing H_2O_2 to H_2O; (4) Fe coordinated to N ($Fe-N_x$) serves as an active site for $4e^-$ direct reduction of O_2 to H_2O. The ratio of the amount of pyridinic and $Fe-N_x$ to the amount of pyrrolic nitrogen serves as a rational design metrics of M-N-C electrocatalytic activity in oxygen reduction reaction occurring through the preferred $4e^-$ reduction to H_2O.

There are two main hypotheses in the literature: one claiming that nitrogen functionalities on/in carbon-based support are directly responsible for their ORR activity and the second suggesting that nitrogen groups serve as the coordinating environment for metal ions, serving as reactive centers for the ORR. Theoretical calculations have confirmed that incorporation of nitrogen atoms in the carbon matrix enhances its electronic properties, but nitrogen atoms are not catalytically active towards a direct reduction of oxygen. Our mechanistic and structural studies have shown that multiple (yet similar) chemical functionalities most likely co-exists in practical M-N-C catalysts, displaying engagement in various steps of the ORR and thus obscuring the "simple model" of the "most active "***active site***".

471580 Electrocatalytic Generation of H_2O_2: Carbon Based Material Synthesis and Device Design for Portable Low Cost Water Purification

Abstract:

Hydrogen peroxide, produced at a rate of 3 million tons/year, is one of the most important industrial chemicals widely used as a bleaching and oxidation agent.[1] Compared to chlorine which is the most commonly used water purification agent at the moment, hydrogen peroxide does not generate harmful byproducts after decomposition, and the higher oxidation potential makes it a more favourable reagent for treatment of water pollutants.[2] Hydrogen peroxide is mainly produced in large-scale facilities using a batch anthraquinone oxidation process, facing the challenge of high transportation and capital cost as well as explosive hazard during storage. For further applications in developing world a selective electrochemical catalyst towards H_2O_2 production is important and a prototype fuel cell/electrolyzer-solar cell device needs to be designed to enable continuous, small-scale and decentralized production of H_2O_2.

Through DFT calculation, we have predicted that different types of nitrogen-doped (N-doped) carbon or carbon with defects are able to selectively catalyze a 2-electron transfer oxygen reduction. We synthesized different types of carbon based materials and tested them along with some commercially available mesoporous carbons. Our results suggest that carbon defect sites without heteroatom doping could serve as the active center. The best catalyst by far is able to catalyse the reaction with no over-potential under alkaline condition with over 90% selectivity.

In this study, prototype devices, both fuel-cell and electrolyzer setups have also been designed and tested. Early results in a fuel-cell setup shows that H_2O_2 has higher affinity towards the aqueous phase as compared to the gas phase. Also H_2O_2 can be easily decomposed in contact with transition metals cations and reductive organic compounds. Considering the fact that electrocatalytic H_2O_2production will have the largest impact in the developing world, itâ€™s desirable to have a simple material manufacturing process and low material cost. All these considerations lead to the current electrolyzer design using plastic with no Nafion membrane and precious metal catalysts involved.

In our electrolyzer design, the produced H2O2 is up-concentrated in the electrolyte. At the current stage, the electrolyzer cell is able to operate with 50 mA overall current at pH 13 with an applied cell potential 1.7 V and faradaic efficiency close to 100%. The device can also be connected to a PV system as the energy supply. A ten-minute operation will able to accumulate 10ml solution with the concentration around 500mg/L.

1. Campos-Martin, J. M.; Blanco-Brieva, G.; Fierro, J. L. G., Hydrogen peroxide synthesis: An outlook beyond the anthraquinone process. *Angew Chem Int Edit* **2006,** *45*(42), 6962-6984.

2. Chu, S.; Majumdar, A., Opportunities and challenges for a sustainable energy future. *Nature* **2012,** *488* (7411), 294-303.

473513 Niobium Doped Molybdenum Disulfide Catalyst for CO_2 Reduction Reaction

Abstract:

The world population consumes an average of 15 trillion watts of power, 85% of which comes from burning fossil resources including petroleum, natural gas, coal, and wood.[1] Despite their low cost and high quality, the use of fossil fuels is unavoidably coupled to the release of many compounds such as carbon dioxide (CO_2) that may adversely affect the environment such as acidifying oceans and climate change.[2,3] Thus, controlling CO_2 emission is essential using either renewable energy sources or remediation techniques. Among various remediation techniques, electrochemical reduction of CO_2 is known as the best approach since it not only reduces the amount of existing CO_2 in the atmosphere but also recycle the inert CO_2 molecule into energy-rich products (e.g., syngas). Layered transition metal dichalcogenides (TMDCs) have recently shown a significant potential to use as the catalyst for a wide range of electrochemical energy conversion and storage systems such as hydrogen evolution and CO_2 reduction[4,5]. However, it has been always a challenge to tailor the electronic structure of TMDCs to achieve highly promoted catalysts.

In this report, we have discovered a superior catalytic performance of Niobium (Nb) doped vertically aligned MoS_2 (VA-MoS_2) as an energy efficient and highly active catalyst for syngas production. Our results indicate that with only 5% Nb concentration at the edge atom structures, MoS_2 exhibits more than two orders of magnitude higher activity than Ag nanoparticles and one order of magnitude higher than pristine MoS_2 at low overpotentials (-0.2 V vs RHE).[6] Additionally, the product characterization results reveal that this promoted catalyst is selective for syngas (CO and H_2) formation at the potential range of -0.164 V to -0.764 V vs RHE. The effect of doping on the structure of pristine MoS_2 was studied by synthesis and testing

different concentrations of dopant for the CO_2 reduction reaction. Different characterization methods such as scanning transmission electron microscopy (STEM), X-ray photoelectron spectroscopy (XPS), electron energy loss spectra (EELS), energy-dispersive X-ray spectroscopy (EDX) and in-situ differential electrochemical mass spectrometry (DEMS) were applied to fully elucidate the effect of dopant on the electronic structure of MoS_2 and its catalytic activity and selectivity.

References:

(1) Davis, S. J.; Caldeira, K.; Matthews, H. D. Future CO2 Emissions and Climate Change from Existing Energy Infrastructure. *Science.* **2010**, *329*, 1330–1333.

(2) Haszeldine, R. S. Carbon Capture and Storage: How Green Can Black Be? *Science.* **2009**, *325*, 1647–1652.

(3) Lastoskie, C. Caging Carbon Dioxide. *Science.* **2010**, *330*, 595–596.

(4) Behranginia, A.; Asadi, M.; Liu, C.; Yasaei, P.; Kumar, B.; Phillips, P.; Foroozan, T.; Waranius, J. C.; Kim, K.; Abiade, J.; *et al.* Highly Efficient Hydrogen Evolution Reaction Using Crystalline Layered Three-Dimensional Molybdenum Disulfides Grown on Graphene Film. *Chem. Mater.* **2016**, *28*, 549–555.

(5) Asadi, M.; Kumar, B.; Liu, C.; Phillips, P.; Yasaei, P.; Behranginia, A.; Zapol, P.; Klie, R. F.; Curtiss, L. A.; Salehi-Khojin, A. Cathode Based on Molybdenum Disulfide Nanoflakes for Lithium–Oxygen Batteries. *ACS Nano,* **2016**, 10 (2), pp 2167–2175.

(6) Asadi, M.; Kumar, B.; Behranginia, A.; Rosen, B. a; Baskin, A.; Repnin, N.; Pisasale, D.; Phillips, P.; Zhu, W.; Haasch, R.; *et al.* Robust Carbon Dioxide Reduction on Molybdenum Disulphide Edges. *Nat. Com.* **2014**, *5*, 4470.

451824 Technology Performance and Economical Feasibility of Bioethanol Production in Pervaporation Membrane Bioreactor with Mechanical Vapor Compression

Abstract:

A pervaporation membrane bioreactor with mechanical vapor compression was developed for bioethanol production. During the experiments of ethanol fermentation-pervaporation process, ethanol could be removed *in situ* from the broth and ethanol production could be enhanced with ethanol inhibition relieved. The yeast cell could continuously grow for long time with cell density of 38 g/L obtained. The byproducts accumulation in the broth led to the culture deterioration of the fermentation and became the main inhibitor against the cell growth. At the end of the fermentation process, over 80 g/L of the byproducts were determined in the broth and this concentration could indeed inhibit ethanol fermentation. Increasing the pH could decrease the concentration of the undissociated weak acids in the broth and improve the fermentation. During the pervaporation process, the total flux of the polydimethylsiloxane (PDMS) membrane was in the range of 350 g m^{-2} h^{-1} and 600 g m^{-2} h^{-1}. At the downstream of the membrane, part of the permeate vapor under the vacuum condition before the inlet of the vacuum pump was condensed at the first condenser by the running water at the room temperature and the non-condensed vapor enriched with ethanol was compressed to the atmospheric pressure and pumped into the second condenser by the vacuum pump. The vapor in the second condenser was easily condensed into a liquid by air at room temperature since the pressure was increased. Ethanol concentration of over 50wt% could be obtained in the second condenser, which could enhance the process separation factor. Energy integration could be achieved by the use of mechanical vapor compression heat pump and the Coefficient Of Performance (COP) of the heat pump could reach 7-9. Improved separation factor and energy integration increased the energy efficiency of the pervaporation process. The cost of ethanol produced in the pervaporation membrane bioreactor was lower, compared with that in the traditional batch fermentation or coupled process with cold trap condensation.

459562 Numerical Simulation of Downdraft Biomass Gasifier: Influence of Structure Designs Using CFD Simulation

Abstract:

Gasification is becoming a more and more popular process for biomass thermal conversion because of less toxic gas and more valuable gas production. However, conducting experiment usually costs a lot of time and resources. Thus numerical study approaches such as CFD (Computational Fluid Dynamics) provide efficient alternative ways for structure design to reach optimal operations.

In this work, three-dimensional CFD models were proposed to simulate flow fields, temperature profiles and gas composition distributions of downdraft biomass gasifiers. At first, one validation model was proposed according to experimental works to compare gas compositions and temperature distributions with real data. It was found that the simulated results were comparable with experimental results. After validation, gasifier structures were modified for optimal reactor designs. Parameters such as air injection positions, air injection angles were studied for better mixing of air with pyrolysed products. Changes in temperature profiles were also studied for fully utilization of the gasifier with high converting efficiency. Also, the influence of reduction-cone shapes was also studied for high quality gas production and temperature distribution. The objective of this study is to use numerical methods to provide guidance for real gasifier design and optimization.

463594 Investigations on the Mass Diffusion Across Interfaces of Oscillating Bubbles

Abstract:

The mass diffusion across interfaces of oscillating bubbles is a complex phenomenon, leading to great difficulties for the numerical investigation. During nonlinear oscillations of bubbles, the position of the interface moves rapidly. Hence, the concentration of the gas (or vapor) in the bubble varies significantly. The fast moving interface also generates a strong velocity field, which prominently affects the local mass diffusion process. If the bubble contains more than one species (e.g. a mixture bubble with both gas and vapor or bubbles with different types of gases), there will be more paramount influencing factors involved. In the present paper, a new numerical approach is developed to simulate the mass diffusion problem. In our model, liquid compressibility, surface tension and nonlinearity are all included. Plus, we further considered the complexity caused by the mixture bubbles with both gas and vapor inside. The results are demonstrated in terms of the threshold of the diffusion (corresponding to the zero flux rate) and diffusion rate for a given set of parameters. The potential application of the main findings are also explained with the background of industrial cases.

472796 Modeling and Simulation of Transport Effects to a Single Reactor Pellet

Abstract:

The purpose of this study is the development and simulation of a mathematical model describing species reaction and transport, as well as energy transport at the pellet-scale. Mass, momentum, and energy are continuously interchanged between the gas phase surrounding the pellets and the catalyst pelletsâ€™ gas and solid interior. The velocity, temperature, and speciesâ€™ concentration profiles within the catalyst pellet can be captured by momentum/energy/species transport models accounting for convection/conduction/reaction/diffusion mechanisms. The dusty gas model is combined with speciesâ€™ and energy transport equations to derive the pellet-domain model. Investigation of pressure effects, pellet size, and pore distribution on species transport along the pellet radius is conducted. The effectiveness factor of each species is calculated.

472800 Effect of Reactor Height on Intensification of Mass Transfer Coefficients ($k_L a$, k_L) in a Turbulent Inverse Bubble Column

Abstract:

In many gas–liquid contactors or reactors, mass transfer from the gas phase to the liquid phase is a major player in the reactor's overall production and/or economics. This talk presents chemical and hydrodynamic data from research to intensify gas-liquid mass transfer in an inverse bubble column bioreactor meant to upgrade natural gas. We measured $k_L a$ by dynamic oxygen absorption and turbulence properties by high-speed camera particle imaging velocimetry (PIV) for four reactor heights. Physical understanding is sought by linear addition of previous theories of k_L under turbulent conditions (Calderbank & Moo-Young 1961) or bubble terminal rise conditions (Clift 1978) to determine $k_L a$ and turbulence intensity at various locations in the reactor. We also add entrance effects according to Higbie's "penetration theory" for interfacial transfer, wherein the freshly created gas-liquid interface leads to higher absorption rate than the classic Calderbank or Froessling type of equations. However, this entrance region transfer rate does not persist long enough (or cover a large enough volume fraction of the reactor) to be dominant over the mass transfer that is forced by other phenomena such as turbulence or bubble-rise-velocity. We observed $k_L a$ values to decrease as the height of the reactor increased from 31 to 124 cm. A model was constructed in which k_L is a sum of effects from bubble terminal rise velocity, turbulence and entrance effects. As a typical bubble moves from jet region (high turbulence) to the bottom of the reactor (low turbulence), energy dissipation decreases, which leads to a decrease in $k_L a$ value. The modeled data match experimental data well. Based upon the results we hypothesize that the best method for intensification of $k_L a$ is by increase of a because the effect of turbulence on $k_L a$ is muted by *a* ¼ exponent. Turbulence and bubble rise velocity is found to mediate k_L in most of the reactor.

References:

Calderbank, P.H. & Moo-Young, M.B. 1961 The continuous phase heat and mass transfer properties of dispersion. Chem. Engng Sci. 16, 39-54.

Clift, A., Grace, J.R. & Weber, M.E. 1978 Bubbles, Drops, and Particles. Academic Press, Inc. New York.

Higbie, R. 1935 The rate of absorption of a pure gas into a still liquid during short periods of exposure. Transactions of the American Institute of Chemical Engineers. 31, 365.

474599 Heat and Mass Transfer Enhancement in a Packed Bed, Low Temperature Fischer-Tropsch Reaction Via Ring-Disc Passive Flow Disturbers

Abstract:

Heat and Mass Transfer Enhancement in a Packed Bed, Low Temperature Fischer-Tropsch Reaction via Ring-Disc Passive Flow Disturbers.

Steven Wiryadinata[1], and Paul Erickson[1]

[1] Energy Research Laboratory, Department of Mechanical and Aerospace Engineering, University of California, Davis

This work investigates the effects of ring-disc flow disturbers on conversion and C_{5+} selectivity of the low temperature Fischer-Tropsch (FT) reaction in a packed-bed tubular reactor with an aspect ratio (length to inner diameter, L/D) of 52. The flow disturbers directly influence the flow characteristics, temperature distribution and species concentrations within the bed, which, in turn, affects the distribution of products (ranging from CH_4 to heavy hydrocarbon waxes and water) from the conversion of reactant syngas (H_2 and CO). The packed bed for FT synthesis is known to suffer from severe radial and axial heat limitation manifesting as large temperature gradients which lead to lower C_{10+} selectivity and increased risk of catalyst degradation. Even though the packed bed affords some flexibility in the reduction of catalyst size to minimize the mass transfer resistance and improve conversion, it is limited by pressure drop considerations, and by the catalyst attrition specifically in terms of reduced structural strength as particle size decreases.

Industrial scale packed bed FT reactor has employed high aspect ratio reactors (L/D in excess of 150) in order to achieve acceptable conversion levels, and significant product liquid recycle in order to remove the heat of reaction [1]. Other smaller scale experimental studies has demonstrated the benefits to conversion and selectivity of inter-stage hydrocarbon products and H_2O removal to maintain high reactant concentration at each reaction state[2], and of the use of helical mixers to promote mixing and enhance the radial heat transfer[3]. The slurry bubble column [4] and the microchannel reactors [5] are alternate reactor designs which address the heat transfer limitation of the conventional packed bed, but introduces other issues such as product-catalyst separation especially with finer particles to reduce mass transfer resistances, catalyst inventory in reaction channel with small hydraulic diameter, and manufacturing and operational cost. With the highest catalyst inventory of all reactor types, the conventional packed bed has the highest productivity but only if heat and mass transfer resistances are minimized.

The ring-disc flow disturbers change the flow characteristics within the bed by inducing mixing and redirecting flow radially to minimize axial channeling. Radial heat transfer is no longer dependent solely on diffusion, and heavy hydrocarbons formed in the vicinity of the relatively cool wall region are directed to the warmer center axis to assist in the heat removal process. For a given inlet mass flow, the rings and discs induce a higher local flow velocity which increases both the convection heat transfer and the external mass transfer.

The experimental setup consists of feed gas conditioning systems, tubular reactor with external cooling jacket, two-stage liquid trap with condenser, and product analysis subsystems. A reactant stream is formulated from bottled gases in 2:1 H_2/CO ratio and heated to 210^oC. A 2.1 cm diameter stainless steel tube houses 15 wt% Co/Al_2O_3 catalysts in pellet form (average diameter of 1 cm) with varying numbers of disturber packages (up to 8 alternating ring-disc packages) in equidistant packing. Saturated propylene glycol maintains the reactor wall at 200^oC. A pressure transmitter and back pressure regulators downstream of the hot trap maintain the reactor pressure at 5 bar. The hot trap is operated at a minimum temperature of 150 ^oC. The effluent line is heat traced to maintain the specified temperature. Hot trap effluents are expanded to atmospheric pressure and cooled in a condenser at 2^oC. The hot trap vapor is directed to a 2nd trap after cooling to separate non-condensable species for online gas analysis of H_2, CO, CO_2 and CH_4. An SRI gas chromatography unit with a 13X molecular sieve and MXT-1 capillary column was used to speciate condensed products.

The flow disturber technique presented here is expected to improve productivity of the packed bed by reducing heat and mass transfer resistances. Optimization of the ring-disc geometry, reactor aspect ratio and catalyst size has the potential of further performance improvements.

ADDIN EN.REFLIST 1. Dry, M.E., The Fischer-Tropsch process: 1950-2000. Catalysis Today, 2002. 71: p. 15.

2. Dai, X.P., et al., Fischer-Tropsch synthesis in a bench-scale two-stage multitubular fixed-bed reactor: Simulation and enhancement in conversion and diesel selectivity. Chemical Engineering Science, 2014. 105: p. 11.

3. Narataruksa, P., et al., Conversion enhancement of tubular fixed-bed reactor for Fischer-Tropsch synthesis using static mixer. Journal of Natural Gas Chemistry, 2012. 21: p. 10.

4. Krishna, R. and S.T. Sie, Design and scale-up of the Fischer-Tropsch bubble column slurry reactor. Fuel Processing Technology, 2000. 64: p. 33.

5. LeViness, S., et al., Velocys Fischer-Tropsch Synthesis Technology - New Advances on State-of-the-Art. Top Catalysis, 2013. 57: p. 8.

452597 Simulation of Toluene Emissions Biodegradation in Bubble Column By Computational Mass Transfer

Abstract:

A computational mass transfer (CMT) model for simulating the bioreaction process in bubble column is proposed. In the proposed model, the conventional method of k-ε model is used for closing the momentum equation for the fluid dynamics; while for the closure of the turbulent mass transfer equation, the recently developed models of concentration variance c^2 and its dissipation rate ε_c are adopted, thus the turbulent mass transfer diffusivity can be determined without assuming the turbulent Schmidt number or using the experimental measured dispersion coefficient as did in a traditional method. With the proposed model, the species concentration distribution and turbulent mass diffusivity distribution along the bubble column can be predicted. To validate the proposed model, simulation is carried out for the biodegradation process of toluene emissions in a bubble column. The simulation results are compared with the experimental data in literature and satisfactory agreement is found between them. Furthermore, the simulations reveal that the turbulent mass transfer diffusivity varies along axial and radial directions, and the turbulent Schmidt number is not a constant throughout the bubble column. Thus, comparing to the traditional method with constant turbulent Schmidt number, the proposed method can fundamentally guarantee more reliable results on concentration distribution in the whole bubble column.

474550 Session Keynote - Practical Membranes Based on Heteropoly Acds for Hotter and Drier Automotive Fuel Cells

Abstract:

The proton exchange membrane (PEM) fuel cell is currently being commercialized for automotive applications, but there is a still a need for PEMs that operate at higher and drier operating conditions and for cathode electrode solution that use less platinum for the oxygen reduction reaction. The heteropoly acids (HPAs) are a subset of the polyoxometalates that have very high acidity in the solid state and have interesting interactions with colloidal platinum group metals. We have demonstrated that very high proton conductivity can be achieved in a model system in which we co-polymerized a divinyl substituted lacunary HPA with butyl acrylate and a small amount of hexanedioldiacrylate (for additional cross linking). While this chemistry had been demonstrated before we were the first group to fabricate freestanding films in the acidic proton conducting form. Proton conductivities >0.1 S cm^{-1} at <90°C and an RH of 50% where demonstrated with this system. Unfortunately, acrylate chemistry is not applicable to practical fuel cells, as the ester linkage in the polymer backbone is not stable to hydrolysis and the methylene units in the polymer are subject to oxidative degradation. We are now developing fuel cell ready hybrid HPA membranes based on perflourinated backbone polymer chemistry. We use dehydrofluorination of a commercial perfluorinated polymer to functionalize the material to allow binding of a lacunary HPA. The emphasis in this project is on film processing to achieve films with nano-scale phase separation and thicknesses of <20 Î¼m. When reproducible film chemistries and processing have been achieved, proton conduction mechanisms are elucidated by pulsed field gradient NMR with EIS and explained based on film morphology elucidated through SAXS and microscopy. The results from these fundamental transport studies help to support simulation studies using Multistate-empirical valance bond theory to explain the exact nature of the proton transport mechanism. One intriguing aspect of the work is that we can use HPA functionalized carbons in the catalysts layer, which should lower the resistance of the membrane electrode interface, while stabilizing the Pt catalysts particles. Recent materials have high enough performance that we are incorporating them into membrane electrode assemblies for single cell fuel cell studies.

475288 Session Keynote - New Designs for Anion Conducting Membranes

Abstract:

Membranes serve important functions as ion and water transport and separation media in water purification, energy storage and conversion systems, and food processing applications. We have explored the factors that control the ionic resistance and permselectivity of anion exchange membranes in a number of polymeric architectures. To decrease the resistance of these materials we have employed new macromolecular synthesis strategies including inducing nanophase separation in the material by long pendant chains. Appending hydrophobic linkers to the polar polymer backbone induces hydrophilic/hydrophobic nanophase separation to boost ion conductivity in the material. Additionally, the mechanical properties of these membranes have been enhanced by forming a semi-interpenetrating network between a poly(ethylene oxide) flexible additive with the cationic poly(phenylene oxide) base polymer. Finally, we have demonstrated patterned membranes through a unique 3D printing process for rapid fabrication of surface-patterned, profiled membranes. This process allows for rapid iteration of new membrane designs for lowing the resistance and improves the hydrodynamics of membrane processes. These innovations in anion-conducting membranes will be discussed along with their performance in devices.

475289 Session Keynote - Alkaline Conducting Membranes and Components for Electrochemical Devices

Abstract:

We developed and/or deployed anion- (mostly hydroxide) conducting materials for several applications. These applications include metal-air batteries, alkaline fuel cells, redox flow batteries and electrochemical reactors of various types. The different environments in which the membranes operate entail different requirements. We will provide an overview of the membrane behavior in these applications to illustrate critical properties. In addition, we will discuss our studies of various electrocatalysts for use in alkaline systems, operated as primary or secondary cells.

474688 Understanding Anion Exchange Membrane Performance in Fuel Cells and Their Shortfalls from Practical Standpoint

Abstract:

Low temperature polymer electrolyte fuel cells can be divided mainly into two categories, proton exchange membrane fuel cells (PEMFC) and anion exchange membrane fuel cells (AEMFC). Anion exchange membranes (AEMs) are an integral part of a working AEMFC. AEMs act as a medium for transporting hydroxide ions from cathode to anode while serving as a separator for fuel and oxidant.

We have explored block and random copolymer morphologies for maximizing hydroxide conductivities and mechanical strength in this work. Polyethylene and Poly(phenylene oxide) based block copolymer AEMs were characterized for their ion transport properties and presence of different water environments. PPO based block copolymers have shown > 100 mS/cm hydroxide conductivity and increased stability which is very promising for AEMFC applications. SAXS, FTIR and diffusion NMR were used to understand membrane morphology and correlate different types of water environments present in the AEM.

Scale up of these synthetic polymers has been an issue faced by many AEMFC researchers for fuel cell performance testing. In this work we have taken an approach of synthesizing commercially available polymers to make 10â€™s of grams of polymer batches which will be sufficient for studying fuel cell performance. We have decided to work with commercially available polyphenelyne based polymer namely poly(2,6-dimethyl-1,4-phenylene oxide). We have functionalized this polymer with quaternary ammonium cations by adopting a chlorination chemistry followed by solvent casting and melt pressing to form a strong AEM. This AEM was used for fuel cell performance testing. Fuel cell performance testing was done on this AEM using H_2/O_2 with Pt/C catalyst at 60Â°C and 80Â°C and currently we are exploring various AEM casting and post synthesis modification techniques for maximizing the fuel cell performance. We think that these post synthesis modifications are critical for optimizing the physical and chemical microenvironments in the AEMs.

Overall this talk will focus on understanding ion transport in novel AEMs and their characterization, followed by fuel cell performance testing and optimization.

Acknowledgements:

We would like to thank the US Army Research Office for funding this Multidisciplinary University Research Initiative under contract #W911NF-10-1-0520

469534 In-Line Quality Control of Proton Exchange Membranes for Fuel Cells

Abstract:

Fuel cells have been in development since the 1960s and stand on the cusp of commercialization for large-scale applications such as non-polluting vehicles. However, they are held back by high manufacturing costs and expensive catalysts. The membrane alone accounts for as much as 45% of the total cost of a commercial fuel cell system. Moreover, manufacturing defects in the membrane lead to scrapped materials and cause cell failures that can cascade into complete stack failure, requiring additional time reworking the stack. Mainstream Engineering is developing an optical sensor system that uses cross-polarized light, providing significant benefits in increased resolution, improved accuracy, and increased detection speeds for the examination of fuel cell membranes and other films. The inspection technique samples every location on the roll stock such that defects (e.g., pinholes, scratches, bubbles) in membrane electrode assembly materials can be removed prior to assembly into fuel cell stacks and end-use devices, such as vehicles. Two different measurement principles for thickness determination using the same hardware are under evaluation. A prototype system was designed and integrated into a 14-inch-wide demonstration web line and was tested running autonomously in real time with PFSA membrane at line speeds up to 100 ft/min. A variety of other fuel cell membranes were also tested in other static or small-scale test configurations. Optical sensor development and pilot testing as part of a manufacturing quality control strategy will be presented.

469262 *The Effect of Imidazolium Ionic Liquids on the Properties of Sulfonated Poly (styrene-isobutylene-styrene) Membranes*

Abstract:

Ionic liquids (ILs) with different anions and cations were incorporated in sulfonated poly(styrene-isobutylene-styrene) (SIBS) as an attempt to modify itâ€™s chemical, thermal, morphological, and transport properties. Different amounts of ILs and different solvents were also studied in order to have a better understanding of the incorporation process and the ability of the solvent to enhance the interaction between the IL and the sulfonated polymer. Membranesâ€™ morphological characterization was performed with small and wide angle X-ray scattering (SAXS and WAXS). Significant variations were observed when different concentrations of ILs were used. The effect of the different chemical variations and compositions of ILs were analyzed with FT-IR to evaluate chemical interactions and with AFM characterizations to evaluate surface effects and morphology changes. The most significant changes were obtained when studying the effect of concentration of the same IL. Phase images showed changing domain sizes as the IL concentration changed. Thermal and transport properties were studied and the results from the morphological and chemical characterizations were used to explain changes in these properties.

472327 Effect of Block Composition on the Morphology, Hydration and Transport Properties of Sulfonated PS-b-Pegpem-b-PS

Abstract:

This work discusses the effect of block composition on the properties of proton conducting polymer membranes. The starting block contained poly (ethylene glycol phenyl ether methacrylate) bifunctional macroinitiator. Polystyrene was added to both sides of the block using atom transfer radical polymerization. Two different percentages of polystyrene (18% and 31%) were synthesized. These copolymers and the homopolymer were completely sulfonated. The resulting polymers produced different water absorption values and transport properties for direct methanol fuel cell (DMFC) applications. The nanostructure and morphology of the casted membranes were studied using small angle x-ray scattering and atomic force microscopy. The results revealed that all six membranes exhibited a disordered phase-segregated morphology, which changed upon sulfonation into small-interconnected ionic domains. Normalized DMFC selectivities (proton conductivity over methanol permeability divided by the respective values for NafionÒ) were calculated and ranged from 1.16 to 15.30, indicating that the performance of these materials can be comparable or better than NafionÒ. Transport property results also suggest that chemistry (block nature and composition), morphology and water content play a critical role in the transport mechanism of protons and methanol. For example, the block copolymer with 18% percent polystyrene displays shorter interstitial ionic distances and sufficient water content to produce high proton conductivity, while maintaining low methanol permeability in a multi-ionic proton exchange membrane.

474701 Session Keynote - Integration, Control, and Optimization of Hydrogen Energy Storage in Utility Grid Networks

Abstract:

Global warming, air pollution, and energy security concerns are evolving the way energy is delivered and converted over the next century with huge increases in the levels of clean and renewable energy while reducing energy demand in all economic sectors. Renewable power is the first and foremost option for decarbonizing electricity generation since it is abundant, sustainable, has lesser emissions and environmental impacts, and is widely available globally. It has been shown that multiple combinations of renewable resources including solar, wind, geothermal, hydropower, and biopower are capable of meeting 80% of total U.S. electricity demand in 2050 while achieving deep reductions in electric sector emissions. However, greater levels of renewable power use pose new challenges to electric grids due to geographical dependency, transmission constraints and most importantly power output variability. Therefore, increasing penetrations of renewable power must be accompanied by higher flexibility in the electricity grid through a portfolio of supply- and demand- side technologies, including energy storage, transmission and distribution infrastructure expansion, more highly responsive loads, and enhanced power system operation.

A compressed hydrogen energy storage system, comprised of an electrolyzer that produces hydrogen from water by using otherwise curtailed renewable electricity, and a fuel cell that utilizes the hydrogen to provide power to the grid, has been considered less favorable than batteries due to lower round-trip efficiency and relatively high cost. However, for integration with large-scale wind energy, accomplishing massive energy storage, and for storing over long periods of time (e.g., seasonal), large energy capacity and low self-discharge become more important than round trip efficiency; therefore such hydrogen energy storage systems become more attractive.

A dynamic model to simulate the integration of hydrogen energy storage systems in utility grid networks has been developed to evaluate the potential of hydrogen energy storage in mitigating the impacts of intermittent renewable power on the grid. A reduction in the renewable energy curtailed and the ramping events has been observed under the application of different hydrogen production, storage and power supply strategies. Data from the state of California has been used as a reference for energy consumption and power generation, assuming the evolution of the power market towards the fulfillment of goals set by the different policies and laws of the state. It was found that both diurnal and seasonal load shifting are possible with gaseous storage facilities that are already available and in use for natural gas storage in the state. Moreover, with the optimal management of the storage facilities, it is viable to dispatch the compressed hydrogen, supplying a significant amount of fuel for transportation applications and averting the need for installing additional electrical transmission and distribution equipment.

463243 Frost Control Using Osmotic Membrane Dehumidification

Abstract:

FROST CONTROL USING OSMOTIC MEMBRANE DEHUMIDIFICATION

Frost buildup is a major concern for heat pumps and refrigerator/freezers. Osmotic membrane dehumidification represents an opportunity for controlling frost simply and effectively. The unique application of this technique to heat pumps provides the advantage of independent dehumidification in the cooling mode and frost control in the heating mode.

An osmotic membrane dehumidifier can use a semi-permeable membrane to facilitate capillary condensation of water vapor and the transport of condensed water through the membrane into a salt solution by osmosis. Here a humid gas stream is brought into contact with a semi-permeable membrane, which separates the gas stream from an osmotic (e.g., salt) solution. Some of the pores of the membrane are small enough to permit capillary condensation. Liquid formed within these pores can connect with liquid formed in adjacent pores, collectively forming continuous paths of liquid. These liquid bridges extend across the thickness of the semi-permeable membrane and provide paths by which water can travel across the membrane. Because the membrane is so thin, water concentration gradients across the membrane can be large. This can provide a large driving force for water transport between the humid air and the osmotic fluid. An illustration of this two-step spontaneous process is given below in Fig. 1.

The effectiveness of the membrane, measured as the rate of water removal per unit area of membrane, is key to the compactness and longevity of the dehumidification system. Automotive air conditioning systems have significant constraints on size, while residential, commercial and industrial air conditioning systems are expected to last for generations. High performance membranes can have a major impact on all of these systems, with no significant energy requirements for dehumidification and with reject heat from the vapor compression cooling process used to re-concentrate the osmotic solution, as shown in Fig. 2.

Typically, capillary condensation occurs in the pores of hydrophilic membranes where surface tension attracts and draws condensed vapor along the sides of the pores. **We have now confirmed that hydrophobic pores in a limited size range will enhance both capillary condensation and the transport of condensate out of the pores.** Test results using membranes containing these pores demonstrate uniquely high fluxes of water.

Beckstein and Sansom, in an article in Proceedings of the National Academy of Science demonstrate under condensing conditions that hydrophobic pores will remain void of condensate if they are under a diameter of 0.8 mm and will fill with liquid if over 1.4 nm (Ref1); that is, capillary condensation can happen away from the surface in a large pore. The hydrophobic surface will drive water vapor away from the surface and cause a density gradient that will result in condensation and cluster formation between water molecules. That is typical of biological pores that selectively transport water, aquaporins, which are lined with hydrophobic residues. Kaneko and Iiyama show the importance of cluster-like water assemblies in hydrophobic nanopores using small angle x-ray scattering to elucidate clusters (Ref 2). *Permeation through nanometer pores is rapid because of the reduced friction resulting from repulsion of water from the hydrophobic surfaces.*

Fig. 1

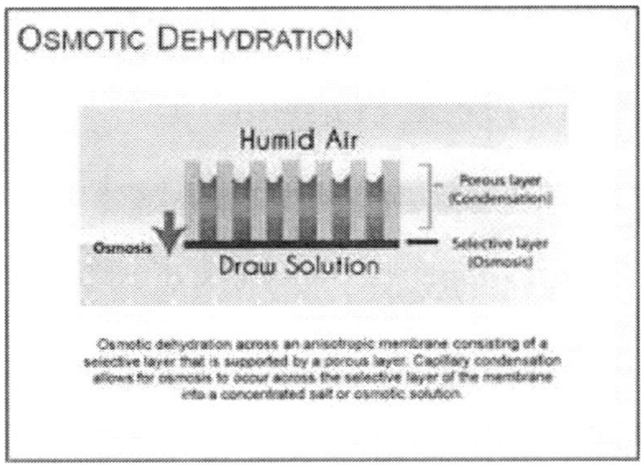

Fig. 2

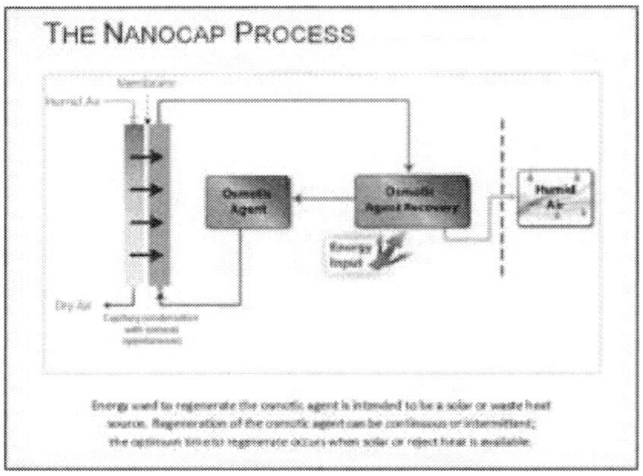

Osmotic dehumidification has been demonstrated in a laboratory scale device using hydrophobic membranes fabricated by Aquaporin Inside, with magnesium chloride used as an osmotic fluid. Water fluxes obtained were more than twice those obtained with available hydrophilic polymeric membranes.

Enhanced performance of the hydrophobic membrane system extends the potential for effective application of osmotic dehumidification to areas like frost control on surfaces containing refrigerant. A heat pump uses a vapor compression system to cool the interior space in warm weather and, by reversing refrigerant flow, to heat the interior space in cold weather. In the cooling mode, independent dehumidification leads to substantial energy savings; in the heating mode, dehumidifying some outside air can inhibit and substantially prevent the formation of frost.

Frost is generally caused by water vapor in very humid air at temperatures somewhat above the freezing point of water freezing on surfaces containing refrigerant at temperature well below the freezing point. Inhibiting the formation of frost involves reducing the humidity of the outside air in contact with cold surfaces and/or raising the temperature of the surfaces in contact with the outdoor air. An osmotic dehumidifier located outdoors in very effective in accomplishing both objectives simultaneously.

As shown in the attached schematic diagram (Nanocap Heat Exchanger.pdf), cold refrigerant exiting the expansion process to the outside is warmed in a liquid-liquid heat exchanger by the osmotic fluid used in the dehumidification process. Outside air that has been dehumidified is transported in a duct to another heat exchanger to warm the refrigerant further. The air is dry enough to eliminate the potential for frost formation, even with refrigerant temperatures below the normal freezing point. The efficiency of the process is enhanced as the temperature is reduced because capillary condensation occurs more readily at low temperature.

Fig. 3

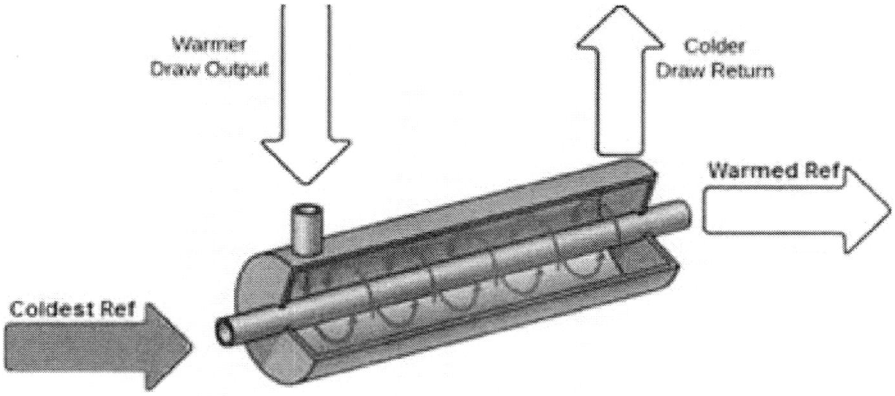

References:

1. Beckstein and Sansom, Liquid-Vapor Oscillation of Water in Hydrophobic Nanopores, Proceedings of the National Academy of Sciences Vol 100, No 12(2003) 7063-7068

2. Kaneko and Iiyama, Structural Understanding of Water Confined in Hydrophobic Nanopores, WATSURF, Les Houches, France Apr. 2013

464623 Optimal Control of a LiBr/Water Absorption Chiller

Abstract:

A LiBr/water absorption refrigeration system driven by solar energy is studied to find its optimal control profile subject to disturbances of the temperature and atmospheric conditions during the day. This work is developed in two phases, first a computational mathematical model was developed to accomplish the optimal control of the chiller, which was then used in a second phase to build a prototype.

A single-stage LiBr/water absorption chiller is made up of four main components: a generator, an absorber, an evaporator and a condenser, which can be modeled as a heat exchanger. The circulation of the solution is assured by a solution pump, and a solution heat exchanger is used to internally recover thermal energy.

Recent works [1-3] had presented dynamic models to evaluate the long-term behavior of a single-effect LiBr/water absorption chiller and also to follow its daily operational profile with good precision. These models may be used to study the interaction of the absorption unit with the other components of a solar-assisted cooling plant, in aim of optimizing both the layout of the system and the control logic of the chiller.

The use of dynamic optimization for analyzing the control possibilities allows to identify the optimal process behavior and select the best control strategy based on quantitative criteria. This implies finding the variable profiles that minimize the objective function in a system subject to disturbances [4]. From the mathematical model or system restrictions a total of 254 variables define the process and a total of 245 equations model the single stage absorption cooling cycle.

The process is modeled by the Differential-Algebraic Equation (DAE) system that represents de dynamics behavior of the system. Then, the optimal control problem is solved by discretizing the time domain, which converts it into a Non-Linear Programming (NLP) problem.

Work [5, 6] results shows that the optimal control strategy is able to optimize the overall performance of the thermal system, achieving a reduction in energy consumption between 0.73 and 2.55% compared with traditional control strategies. With this work we seek to reduce energy consumption for an absorption cycle by applying an optimal control strategy as it has been accomplished for other thermal systems.

Considering that the most inefficient energy consumption systems, identified in countries like Colombia, come from using equipment for cooling and space conditioning [7] and, besides, that during the summer season the energy generation is insufficient to satisfy the demand, the absorption refrigeration system becomes as an opportunity to solve this problem in a more definitive way.

The Non-Linear Programming (NLP) problem is implemented in Gams® and solved with the solver CONOPT.

Reference

[1] G. Evola, N. Pierres, F. Boudehenn, P. Papillon.

[2] P. Kohlenbach, F. Ziegler, "A dynamic simulation model for transient absorption chiller performanc. Part I: The model" International Journal of Refrigeration, pp. 217-225. 2008

[3] P. Kohlenbach, F. Ziegler, "A dynamic simulation model for transient absorption chiller performanc. Part II: Numerical results and experimental verification" International Journal of Refrigeration, pp. 217-225. 2008

[4] J. Cuthrell, L. Biegler, "On the optimization of differential-algebraic process system" Annual AIChE Meeting, Miami, Fl, 1986.

[5] Z. Ma, S. Wang, "Supervisory and optimal control of central chiller plants using simplified adaptive models and genetic algorithm" Applied Energy, pp. 198-211, 2011

[6] T. Chow, G. Zhang, Z. Lin, C. Song, "Gobal optimization of absorption chiller system by genetic algorithm and neural network" Energy and Buildings, pp. 103-109, 2002.

[7] "Plan Energetico Nacional Colombia: Ideario Energético 2050", Unidad de Planeación Minero Energetica, 2015

464689 B-C and B-Fe Core-Shell Nanoparticles for the Enhancement of Biofuel Combustion

Abstract:

Liquid biofuels such as ethanol are good candidates to partially replace fossil fuels. However, the adoption of biofuels is inhibited by low energy densities and incomplete combustion. These characteristics are problematic for applications with a high energy demand, such as ramjets. Boron additives are a promising solution because they are among the most energy-dense materials. Boron can increase the overall heat release of the fuel and reduce its ignition temperature. However, boron gradually oxidizes when stored under ambient conditions, may not completely burn, and the used metal particles lack recyclability and are subject to environmental regulations. These problems are less limiting in ramjets and other specialty combustion applications.

In this study, we synthesized core-shell Boron-Carbon (B-C) and Boron-Iron (B-Fe) nanoparticles for the combustion of ethanol. A variety of templated, non-templated, and wet-chemical techniques were used to produce elemental boron (zero valent) with either a carbon or iron shell. The composites were characterized by N_2 Physisorption, TGA/DSC, ICP/AES, XRD, SEM/EDX, TEM, and SAXS. The composites had surface areas from 20-150 m^2/g, pore sizes from 5-40 nm, and shell thicknesses between 5-20 nm. Boron was well-distributed and mostly amorphous, indicating that reactions with iron or carbon were minimal. The combustion properties with ethanol were determined through batch combustion in a calorimeter. The B-C and B-Fe composites lowered ignition temperatures and raised the rate of fuel conversion in almost all cases. The best B-C composites were synthesized with polymer templates and had relatively large shell thicknesses and pore sizes that positively affected oxygen transport. The best B-Fe composites could be broken down into smaller particles, which is beneficial when fluidized in a ramjet combustor.

474452 Platinum-Nickel Nanowires As Electrocatalysts in Alkaline Hydrogen Oxidation and Evolution

Abstract:

Anion exchange membrane (AEM) fuel cells and electrolyzers have advanced over recent years to become competitive with proton exchange membrane (PEM) devices. The primary benefit of AEM systems is the potential ability to use less expensive, non-platinum (Pt) group metal (PGM) catalysts, and the non-PGM stability that the alkaline environment allows. While these systems have had polymer limitations, recent improvements in the chemical stability of alkaline membranes suggest that catalysts will become the limiting factor in the near future.[1, 2]

Catalyst development in AEM systems typically focuses on oxygen reduction and evolution, since the reactions are kinetically orders of magnitude slower than hydrogen oxidation and evolution. Alternatives, however, exist to PGMs in these reactions: silver in oxygen reduction; and nickel and cobalt in oxygen evolution.[3, 4] While hydrogen oxidation and evolution are kinetically faster reactions, they are roughly two orders of magnitude slower on Pt in base compared to acidic environments. Non-PGM catalyst options are also less clear, and generally struggle to justify the AEM cost benefit, producing activities orders of magnitude lower than PGMs at higher overpotentials.[5]

Recently, advanced Pt electrocatalysts have been developed in an effort to thrift the amount of PGMs in AEM fuel cells and electrolyzers. Pt-nickel (Ni) nanowires, previously developed for acidic oxygen reduction, were studied for their activity in hydrogen oxidation and evolution.[6] These materials were formed by spontaneous galvanic displacement, a process that occurs when a metal template contacts a nobler metal cation. At low levels of displacement, small amounts of Pt were deposited to produce high electrochemical surface areas. Subsequent post-synthesis processing was used to integrate the Pt-rich and Ni-rich zones, compressing the Pt lattice and improving its activity for hydrogen oxidation and evolution. Compared to carbon-supported Pt nanoparticles (Pt/HSC), Pt-Ni nanowires produced hydrogen oxidation/evolution exchange current densities 9 times greater.

[1] B. Pivovar, Alkaline Membrane Fuel Cell Workshop Final Report, in: U.S. Department of Energy (Ed.), http://www1.eere.energy.gov/hydrogenandfuelcells/pdfs/amfc_may2011workshop_report.pdf, 2011.

[2] K.J.T. Noonan, K.M. Hugar, H.A. Kostalik, E.B. Lobkovsky, H.D. Abruña, G.W. Coates, Journal of the American Chemical Society, 134 (2012) 18161-18164.

[3] R. Subbaraman, D. Tripkovic, K.-C. Chang, D. Strmcnik, A.P. Paulikas, P. Hirunsit, M. Chan, J. Greeley, V. Stamenkovic, N.M. Markovic, Nat Mater, 11 (2012) 550-557.

[4] J.S. Spendelow, A. Wieckowski, Physical Chemistry Chemical Physics, 9 (2007) 2654-2675.

[5] J.K. Nørskov, T. Bligaard, A. Logadottir, J.R. Kitchin, J.G. Chen, S. Pandelov, U. Stimming, Journal of The Electrochemical Society, 152 (2005) J23-J26.

[6] S.M. Alia, B.A. Larsen, S. Pylypenko, D.A. Cullen, D.R. Diercks, K.C. Neyerlin, S.S. Kocha, B.S. Pivovar, ACS Catalysis, 4 (2014) 1114-1119.

474645 Characterization of Multicomponent Transport in Membranes for Solar Fuels Devices

Abstract:

Ion conducting membranes are of interest for various energy applications including fuel cells and solar fuel generators. Within the context of artificial photosynthesis towards solar fuel production, membranes are required to facilitate the ion transport necessary to promote electrochemical water splitting and CO2 reduction while meeting additional selectivity and permeability demands. Generally, membranes that limit the crossover of CO2 reduction products are desired, as high rates of product permeation through the membrane typically reduces device efficiency. Herein, we demonstrate the use of in-situ ATR FT-IR spectroscopy to quantitatively resolve the concentration of single component and multicomponent mixtures of select CO2 reduction products; methanol, formate and acetate. We then apply this methodology to the in-situ monitoring of single component and multicomponent permeation across commercially available membranes such as Nafion. Emergent transport phenomena, where the transport of a small molecule solute is affected by the presence of another solute, are explored via multicomponent permeation measurements. Membrane permeabilities and selectivities derived from single component permeation experiments are found to deviate, in some cases significantly, from permeabilities and selectivities measured in multicomponent permeation experiments.

474783 The Evaluation of Ionomers in Alkaline Membrane Electrode Assemblies for Water Electrolysis

Abstract:

The cost of water electrolysis can be reduced by transitioning from an acid to an alkaline environment. The chemistry and electrochemistry of alkaline electrolyte-based electrolyzers allows for less expensive materials-of-construction and catalysts.

The ionomers used to construct the three-dimensional electrodes on the anion conducting membrane (collectively called the membrane electrode assembly (MEA)) contribute to the overall performance of the electrolyzer. First, the ionomer provides the ion pathway between the anion conducting membrane and catalyst particle. Second, the ionomer must solubilize and transport the chemical reactants to the catalyst surface and facilitate the removal of the reaction products. Third, an electrochemical double layer is formed within the ionomer at the catalyst surface. The nature of the double layer affects the overvoltage at the electrode which contributes to the overall cell potential. Fourth, the ionomer provides mechanical stability and adhesion of the particles in the three-dimensional electrode. Thus, the ionomer should have high chemical and mechanical stability so that it does not degrade by the electrochemical or have excess swelling and adversely affect the mass and ion transport to the catalyst.

In this project, a number of ionomers were studied in an alkaline electrolysis cell. The overvoltage and chemical stability of the ionomers was compared. The ionomers tested in this project include those with an all-hydrocarbon backbone, ionomers with partial fluorination in the backbone and ionomers with a block copolymer backbone.

475777 A Fail Safe, Passive Operation Small Modular Nuclear Reactor for Electricity Generation and Process Heat Production

Abstract:

A Fail Safe, Passive Operation Small Modular Nuclear Reactor for Electricity Generation and Process Heat Production Mohamed S. El-Genk

Regents' Professor, Nuclear, Mechanical and Chemical Engineering, and Director, Institute for Space and Nuclear Power Studies, School of Engineering, University of New Mexico, Albuquerque, New Mexico, USA

SMRs are those which provide for a total electrical power generation capacity from as little as 10 to 300 MWe. Depending on the installed electrical power and the type of the SMR plant, all or most reactor components could be built and assembled in the factory and shipped to the construction site by rail, heavy trucks or on barges. The modularity and the short construction time of the plant of 12 – 18 months reduce capital cost and enhances the quality assurance for the reactor and plant components. This flexibility avoids large financial obligations associated with building a large plant, which may require developing and managing a consortium of multiple utilities in more than one state. This is, in addition to paying interest on a large loan ($5B- $8B) for many years and the lack of revenues during 5-6 years construction time.Several countries are pursuing different SMR designs, building prototypes, doing actual construction and planning a near term deployment. These include the Russian Federation, Japan, Argentina, China, India, the Republic of Korea and USA. Nevertheless, there are many challenges associated with developing and deploying SMRs. Examples include: (a) developing a functional central control room for

coordinating multiple operating units on a single site; (b) developing operators training for managing multiple reactor units; (c) licensing multitude of design innovations; (d) reaching an economic competitiveness with large plants and renewable energy sources; (e) establishing future market demand; and (f) developing procedures for the handling, storage and processing of used fuel or reactor unitsThis paper introduces the University of New Mexico's Scalable LIquid Metal cooled small Modular (SLIMM) reactor (Figs. 1 and 2), capable of generating 10 to 100 MW of thermal power, for 55–6 full power year, respectively, without refueling. Natural circulation of in-vessel liquid sodium cools the SLIMM reactor core during nominal operation and after shutdown, with the aid of in-vessel tall chimney and a sodium-sodium (Na-Na), concentric helically coiled tubes heat exchanger (HEX) placed at the top of the down-comer in the reactor vessel.

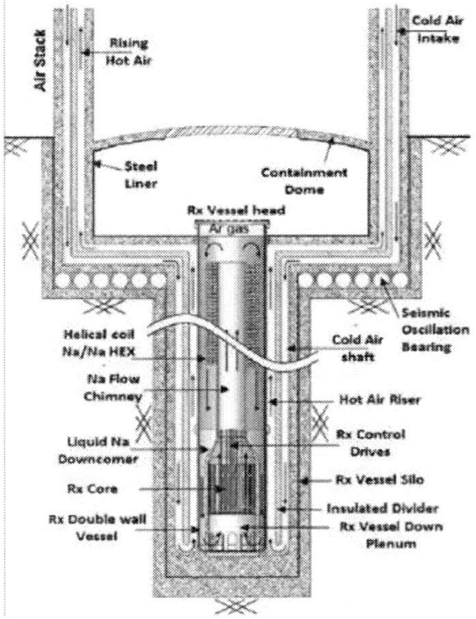

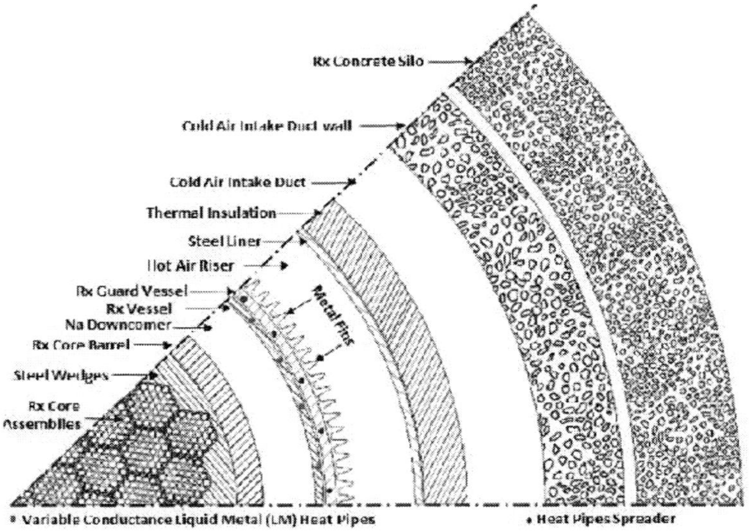

Fig. 2. A Pie Section of the SLIMM reactor Layout.

Fig.1 Axial Cross-sectional View of SLIMM, SMR Installed Below Grade

In case of a malfunction of the HEX, a backup cooling system (BCS) passively removes the decay heat generated in reactor core after shutdown, from the outer surface of the reactor vessel by natural circulation of ambient air (Figs. 1 and 2). This system is capable of removing 1-1.26 MW of thermal power, which together with the large mass of the liquid sodium in the reactor vessel (~ 38 MT), limit its temperature rise after reactor shutdown to < 900 K. Owing to the lower vapor pressure of liquid sodium (boiling temperature is ~ 1156 K at atmospheric pressure), the SLIMM reactor nominally operates slightly below atmospheric pressure. In addition, the high negative temperature reactivity feedback of the SLIMM reactor core would effectively shut it down, in case of a malfunction of the redundant reactor control and emergency shutdown systems of enriched B_4C poison rods.The SLIMM reactor employs UN fuel, and HT-9 steel cladding, core structure and reactor vessel for operating at 600 K inlet and 800 K exit temperatures of the circulating liquid sodium through the reactor during nominal operation. At these temperatures, liquid sodium is compatible with the HT-9 steel with negligible corrosion concerns during the reactor long operation lives. The relatively high sodium exit temperature from the SLIMM reactor core makes it possible to generate high-pressure superheated steam

for achieving a plant thermal efficiency in excess of 40% and provide high temperature heat for a host of industrial uses and the thermochemical production of hydrogen fuel.The SLIMM reactor would be fully fabricated, assembled and sealed in the factory and shipped to the construction site by rail, a truck or on a barge. It would be installed below ground (Fig. 1), to avoid direct impact by missiles or aircraft, and mounted on seismic insulators to resist earthquakes. The variable conductance liquid-metal (LM) heat pipes, along the primary vessel wall (Fig. 2), transport thermal energy to thermoelectric modules to generate 10s of kWe of auxiliary power for operating the plant's vital functions during nominal operation and in case of a total loss of onsite and offsite powers, avoiding a Fukushima Daiichi type disaster.

474730 Session Keynote - Hybrid Renewable Energy Systems: State of the Art of the Impact of Design and Control Strategy on Performance and Costs

Abstract:

The pushing development toward a greater and greater penetration of Renewable Energy Sources (RES), which are characterized by intermittent availability, is giving new challenges through the delocalization of power production, aiming at consuming energy wherever it is originally converted. The use of HRES (Hybrid Renewable Energy Systems), i.e. characterized by the presence of power subsystems generating from renewables (i.e. PV and wind turbine systems primarily), coupled to batteries for short term storage, and Electrolyzers/Fuel Cells to mange storage over the long term, may be interesting to improve the flexibility of the solutions.

Such systems however present some challenges, as the efficiency depends strongly on the control strategy implementation, which in turn may be based on simple rules (RBS=Rule Based Strategy), or on more sophisticated elaborations (e.g stochastic, or model predictive based [1]) taking into account the probable evolution of the system status due to future consumption pattern management (Demand Response) and weather forecasts. Also, the typical small size (in relative terms to usual power plant size) of the systems give rise to a major drive to downsize at the utmost the components and simplify the control strategies (for example minimizing the amount of hardware required to optimize the system behavior, such as inverters, measurements probe, etc) [2].

The solutions are usually evaluated in terms of efficiency and Total Cost of Ownership: it is clear that these parameters must be defined on an energy basis, over a timeframe of at least one year to make sure to take into account the seasonal variation of the weather. It is also very important to highlight that the relation between power production and consumption is very much dependant on the specific application: for example, performances of off-grid systems [3] may be very different from grid-connected [4]; the consumption of energy in electric vehicles [5] or the use of FC as CHP systems may give very different results in terms of cost effectiveness and global efficiency.

The cited aspects will be properly highlighted and commented in this presentation, which will provide the audience with a description of the state of the art of system design, as well as with details on control strategy and main trade-offs.

References

[1] Bruni G., Cordiner S., Mulone V., Rocco V., Spagnolo F,. En Conv Magag., 2015.

[2] Erdinc O., Uzunoglu M., Renew and Sust. Energy

[3] Bruni G., Cordiner S., Mulone V. et al., Int. J. Hydr. Energy, 2014.

[4] Dagdougui H., Minciardi R., Ouammi A., Robba M., Saciel R., En Conv Manag., 2012.

[5] Fazelpour F., Vafaeipour M., Rahbari O., Rosen M.A., En. Conv. Manag., 2014.

475287 Modeling of reverse electrodialysis for waste heat

recovery

Abstract:

Reverse electrodialysis (RED) harnesses salinity gradient energy by controlling the ion transport through ion exchange membranes. Natural resources such as sea and river water can be used in the RED process, thus sodium chloride has been commonly used for a variety of the RED researches. However, those resources require the repetitive shipping and pretreatment cost and can act as a locational limitation for installation of the RED plant. A closed-loop RED, which uses thermolytic solution, can solve these issues as well as generate electricity. In recent years, ammonium bicarbonate (NH4HCO3), which can be separated by relatively low temperature around 60 „*f*, has been investigated as a promising solute that enables the system closed-loop. Therefore, the closed-loop system keeps the salinity gradient by solute separation and recapturing process.
In this study, we developed a numerical NH4HCO3-RED model for the investigation of the RED performance. Because of the unclarified electrochemical information of NH4HCO3 in solution, such as activity and equilibrium constant, the Planck-Henderson equation was used to approximate the membrane potential. And, solution resistance was experimentally calculated by the definition of the molar conductivity. Moreover, permselectivity that has been assumed either as a unity or some constant value in previous researches was also experimentally obtained to estimate the performance more precis ely.
We found the developed NH4HCO3-RED model was in great accordance with the experimental results under various operating conditions. Using the developed model, we also characterized the net power density, which considers the pumping loss. In our system, the maximum net power density of 0.84 W/m2 was obtained at an optimum operating conditions that the intermembrane distance of 0.1 mm, the concentration ratio of 200 (2.0 M/ 0.01 M), and the flow rate of 3 mL/min. We believe that this study will improve our understanding of the NH4HCO3-RED system for& nbsp;a novel waste heat recovery technology.

481270 Techno-economic Analysis (TEA) of Next-Generation Proton-conducting Solid Oxide Fuel Cells (SOFC)

Abstract:

This work develops preliminary techno-economic models for a highly-innovative, next generation proton-conducting, anode-supported solid oxide fuel cell (SOFC) that integrates a methane coupling catalyst into the anode. This cell is under development by Argonne National Laboratory (ANL) and the Illinois Institute of Technology. The cell is currently at technology readiness level (TRL) 1, i.e. proof-of-concept at a laboratory bench scale. This work summarizes (1) a specific physical embodiment of the ANL cell design modelled, (2) the bill of materials (BOM) for this modelled cell, and (3) preliminary cell, stack, and system cost estimates based on this cell. The techno-economic analysis (TEA) focuses on electrolyte-electrode assembly (EEA), or the 'heart' of the electrochemical cell, where the anode and cathode half-reactions occur.

The project methodology is as follows:

1. Develop a physical embodiment of the design for the EEA.
2. Develop a Bill of Materials (BOM) that specifies masses and/or quantities of components and materials in the EEA.
3. Obtain price quotes for materials in the EEA.
4. Estimate materials costs for the EEA to develop a 'price floor' for minimum costs.
5. Identify cost drivers within the EEA.
6. Estimate fuel cell stack costs for a 100 kWe net electric stack mass-produced at a rate of 50,000 systems per year.
7. Estimate fuel cell system costs for a 100 kWe net electric stack mass-produced at a rate of 50,000 systems per year.

Preliminary results indicate that EEA is estimated to cost 5.56 cents/cm^2 or $278/kW of gross electric power from the cell or stack. Within the EEA, the primary cost driver for the EEA appears to be the anode substrate, which is composed of yttrium-doped barium zirconate (BZY or $BaZr_{(1-x)}Y_xO_{3-d}$ or sometimes $BaZr_{0.8}Y_{0.2}O_x$ or $BaZr_{0.9}Y_{0.1}O_{3-d}$), with barium zirconate represented as ($BaZrO_3$), and nickel oxide (NiO). The BZY and NiO anode substrate material is estimated to cost either 2.88 cents/cm^2 [based on price quotes for BZY from TransTech Inc.] or 16.40 cents/cm^2 [based on price quotes from Praxair Specialty Ceramics]. The secondary cost driver for the EEA appears to be the anode's methane coupling catalyst. The methane coupling catalyst is modelled as a dual metal catalyst composed of Platinum (Pt), gallium(III) oxide (Ga_2O_3), and silicon dioxide (SiO_2). The catalyst material is estimated to cost 2.45 cents/cm^2. About 94% of its cost is due to the material cost of platinum as a catalyst.

Preliminary results also indicate that, for a 100 kW net electric stack mass-produced at a rate of 50,000 systems per year, the fuel cell stack subsystem cost is estimated to be ~$478/kW of net electric power from the system. This cost is approximately 74% higher than a 'plain vanilla' SOFC subsystem, analyzed by the author for DOE in a prior analysis. For a 100 kW net electric stack mass-produced at a rate of 50,000 systems per year, the overall fuel cell system cost including all BOP components is estimated to be ~$597/kW of net electric power from the system. This cost is approximately 48% higher than the entire 'plain vanilla' SOFC system, analyzed by the author for DOE in a prior analysis.

452586 An Optimization Approach to Thermodynamic Limitation on Convective Mass Transfer Enhancement

Abstract:

An optimization method to find the thermodynamic limitation on convective mass transfer for fixed viscous dissipation is developed. The method applies the principle of extremum entropy generation in the mass transfer problem. The mass transfer-induced entropy generation in the fluid flow is defined as the criterion. By maximization or minimization of the criterion under a constraint of fixed viscous dissipation, a velocity field and associate body force field can be obtained, which can enhance the mass transfer process to the thermodynamic limitation. By applying the optimization method, the dependence of the maximum extent of convective mass transfer enhancement on the viscous dissipation can be found. Such a limitation for mass transfer is shown to be useful in setting a target for mass transfer enhancement in practical applications. The analysis is illustrated by three examples.

474658 Session Keynote - Hybrid Fuel Cell System for Producing Chemicals and Electricity from Natural Gas

Abstract:

We are developing a hybrid intermediate temperature (500-700°C) proton-conducting fuel cell (PCFC) system that operates directly on natural gas or natural gas liquids to generate electrical power and to produce high value chemicals such as ethylene or propylene as part of a U.S. Department of Energy Advanced Research Projects Agency – Energy (ARPA-E) funded project under its REBELS Program. For natural gas, non-oxidative coupling of methane (NOCM) catalyst technology is integrated into the anode compartment of the fuel cell to produce C_2 and C_3 alkanes, olefins and H_2. For natural gas liquids, propane dehydrogenation (PDH) catalyst technology is integrated into the anode compartment to produce propylene and H_2. The benefit of integrating severely thermodynamically-limited reactions such as NOCM or PDH into a PCFC is that the H_2 produced by these reactions is oxidized within the anode, generating protons which migrate through the electrolyte to the cathode where they react with oxygen to form water. By removing H_2 from the product mixture using the fuel cell, higher conversion levels for these reactions can be achieved at lower temperatures than in a conventional thermal catalytic process (Le Chatelier's Principle). Additional benefits of our hybrid fuel cell system are that the lower operating temperature allows for a wider choice of less-costly materials compared to conventional reactor processes, the use of a proton-conducting electrolyte eliminates the loss of carbon due to the formation of CO and CO_2 that occurs in oxidative coupling of methane (OCM) when oxygen is co-introduced using an oxide ion-conducting electrolyte, and the lower operating temperature minimizes coke formation during propane dehydrogenation due to thermal cracking of propylene.

To develop this hybrid system, we have had to develop (1) a novel proton-conducting fuel cell that is capable of operating efficiently at 500-700°C, (2) NOCM and PDH catalyst technologies that operate with high activity and selectivity at these temperatures, and (3) a methodology for integrating the NOCM or PDH catalysts into the PCFC. This talk will summarize our efforts to develop the proton-conducting fuel cell technology based on a barium-cerium-yttrium electrolyte; the NOCM and PDH catalyst technologies based on single-site catalysts that exhibit high yield and selectivity while minimizing coke formation, methodologies for integrating the catalysts into the anode compartment of the fuel cell, and preliminary test results. Preliminary results from a techno-economic analysis (TEA) comparing of our technology to competing technologies will be presented.

This work is supported by the U.S. Department of Energy, Advanced Research Projects Agency – Energy (ARPA-E) under Contract DE-AC02-06CH11357

474790 A Comparative Techno-Economic Analysis of Renewable Hydrogen Production Using Solar Energy

Abstract:

A technoeconomic analysis of photoelectrochemical (PEC) and photovoltaic-electrolytic (PV-E) solar-hydrogen production of 10,000 kg H2 day-1 (3.65 kilotons per year) was performed to assess the economics of each technology. The systems analyzed include: (i) two PEC systems that are defined by a single unit operation with inputs of solar power and water and outputs of hydrogen and oxygen, and that are differentiated primarily by the extent of solar concentration (unconcentrated and 10x concentrated), (ii) two PV-E systems that are defined by two unit operations, solar panels that convert solar power to electricity and proton exchange membrane (PEM) electrolyzers that convert electricity to hydrogen, and that are differentiated by the degree of grid connectivity (unconnected and grid supplemented), and (iii) PEM electrolyzers supplied by grid electricity to serve as a commercially practiced benchmark.

For each system, a base-case design and bill of materials was developed and used to determine the untaxed, plant-gate levelized cost of hydrogen (LCH). Based solely on laboratory scale demonstrations and thus likely optimistic projections for system costs, the unconcentrated and 10x concentrated PEC systems base-case capital expense and LCH values were $205 MM ($293 per m2 of solar collection area (mS -2), $14.7 W-1 H2) and $11.5 kg-1, and $160 MM ($428 mS -2, $11.5 W-1 H2) and $9.2 kg-1, respectively. Consisting of significantly more mature technologies, the non-grid connected and grid connected PV-E systems base-case capital expense and LCH values were $260 MM ($371 mS -2, $18.8 W-1 H2) and $12.1 kg-1, and $66 MM ($441 mS -2, $4.9 W- 1 H2) and $6.1 kg-1. The benchmark PEM grid electrolysis systems capital expense and LCH were $34 MM ($2.4 W-1 H2) and $5.5 kg-1, respectively. A sensitivity analysis indicated that, relative to the base-case, increases in system efficiency could effect the greatest cost reductions for all systems,

due to the areal dependencies of many of the components. The balance-of-systems (BoS) costs were the largest factor in differentiating the PEC and PV-E systems.

Successful research and development, measured solely by achieving a system efficiency of greater than 20% within the current embodiments of solar H2 generators, is not sufficient to produce systems with LCH values comparable to that of fossil-fuel derived hydrogen. Panel mounting materials, labor, and other balance of systems costs, irrespective of the active materials, amount to LCH values in excess of current hydrogen and energy prices. For basecase PEC systems, a carbon tax of greater than $800 (ton CO2)-1 would be required in an unconstrained CO2 energy market for fossil-fuel derived hydrogen to reach price parity with these solar technologies.

Expected electricity prices from CO2-neutral nuclear and low-CO2 fossil fuel with CCS technologies are currently an order of magnitude lower than expected electricity prices from solar or wind systems with battery or fuel storage solutions that provide reliability similar to that of current base-load or dispatchable generation technologies. Given these low electricity prices, electrolytic hydrogen prices are also projected to be significantly lower than the solar hydrogen routes analyzed, requiring disruptive approaches to solar hydrogen generation relative to the present embodiments of the technology. Efforts to directly reduce CO2 photoelectrochemically or electrochemically could potentially produce products with higher value than hydrogen, but many as yet unmet challenges include catalytic efficiency and selectivity, and CO2 mass transport rates and feedstock cost.

As compared to other, commercial, energy production and storage technologies, the solardriven technologies analyzed herein are limited by an intrinsically diffuse solar power density of <1 kW m-2, as well as by a low capacity factor, of under 25%, for fixed-tilt panels in the optimal locations in the U.S., and by a conversion efficiency of under 20%. The capacity factor of presently known solar-based energy systems is their fundamental limitation; any capital item used only 25% or less of the day will be at a disadvantage to capital used more effectively.
Efforts to increase the low capacity factor of terrestrial solar power, as well as re-designed solar installation schemes that significantly reduce the balance of system costs and increase the efficiency to near the Shockley-Queisser limit without increasing the capital cost, will have the largest impacts on the economic competitiveness of the resulting technology implementations.
Radically new materials and system designs that achieve fully installed costs similar to simple material installations such as artificial grass are required to achieve the equally dramatic cost reductions needed for solar power to compete with current generation technologies.

474724 High Performance LT-SOFC Based on a Strontium Iron Cobalt Molybdenum Oxide Based Ceramic Anode Support

Abstract:

High Performance LT-SOFC Based on a Strontium Iron Cobalt Molybdenum Oxide Based Ceramic Anode Support

Ke-Ji Pan, Colin Gore, Lei Wang, Luis Correa, Thomas Langdo, and Bryan Blackburn

Redox Power Systems, LLC

College Park, MD 20742

Abstract

Solid Oxide Fuel Cells (SOFCs) consisting of a ceramic anode have several advantages when compared to a traditional Ni-cermet based anode. The main advantages include better red-ox cycling stability and chemical stability with the use of hydrocarbon fuels. However, ceramic materials, such as Nb or La doped $SrTiO_3$, are much less conductive than Ni. Furthermore, such materials usually display appreciable conductivity only after a high temperature reduction step, thus practically limiting their use to higher operating temperatures. Unfortunately, the catalytic activity of ceramic anodes is not typically as good as Ni. At low operating temperatures fuel oxidation kinetics are hindered and must be improved through the infiltration of catalysts. In addition, ceramic anode-supported SOFCs have more fabrication difficulties than Ni-cermet cells, such as the potential for chemical reactions between cell components, thermal mismatch, and shrinkage mismatch. Therefore, until now there have been very few reports of ceramic anode-supported LT-SOFCs. In this work, we have successfully synthesized new strontium, iron, cobalt, molybdenum oxide ceramic anode materials with different compositions (SFCM#1-#5) and fabricated anode-supported SOFCs using GDC as the electrolyte. SFCM displays a conductivity greater than 30 $S*cm^{-1}$ between 450 ¡ãC to 650 ¡ãC without the need for a high temperature reduction step, as shown in Figure 1. Button cells have been fabricated and demonstrated a power density greater than 0.4 W/cm^2 at 500 ¡ãC and have shown good stability in H_2/CH_4 mixtures for several hundred hours. We have scaled up the size of the cells to greater than 5 cm by 5 cm (Figure 2) and will discuss additional performance results and future challenges as we work towards integrating the ceramic anode cells into an LT-SOFC stack.

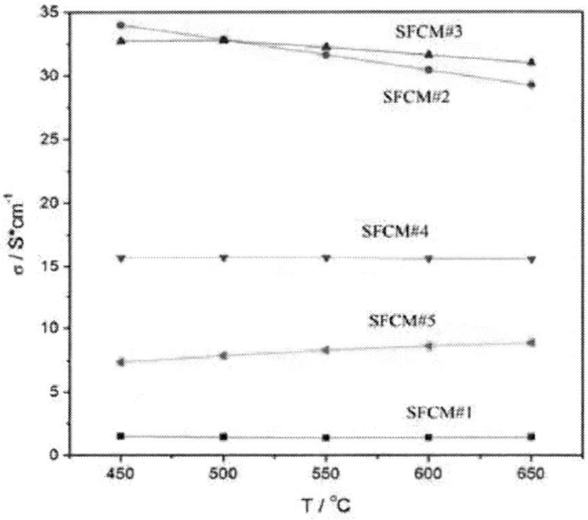

Figure 1. Conductivities of various SFCM compositions in 10% H_2 / 90% N_2

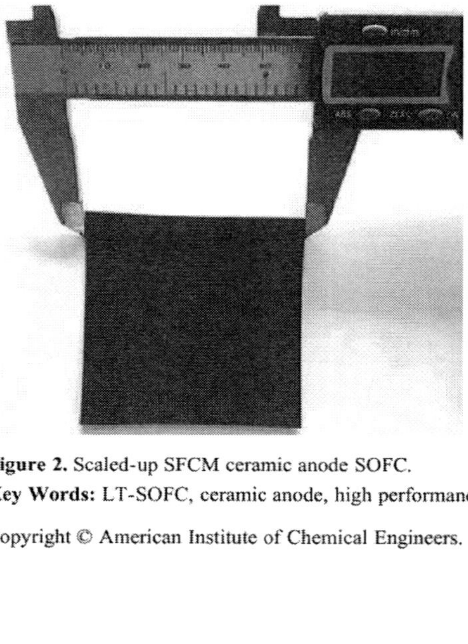

Figure 2. Scaled-up SFCM ceramic anode SOFC.

Key Words: LT-SOFC, ceramic anode, high performance, long term stability

459447 Improved Activity and Stability of Ni-Based Anode for Direct Methanol “Fueled Solid Oxide Fuel Cells

Abstract:

Mo is utilized to modify the Ni- $Ce_{0.8}Sm_{0.2}O_{1.9}$ (SDC) anode. The NiO-MoO_3-SDC anode materials are prepared by impregnation technique, which are reduced to SDC and Ni-Mo alloy phases in H_2 atmosphere. Solid oxide fuel cells (SOFCs) with Ni-Mo-SDC composite anodes are fabricated and tested with methanol as fuel. The addition of Mo improves the catalytic activity for methanol pyrolysis and the resistance to carbon deposition of Ni-SDC anode. The electrochemical performance test demonstrates that the addition of Mo into Niâ€“SDC anode effectively improves performance and stability of the single cell with methanol as fuel at 700 °C. The anode with a mole ratio of Mo to Ni of 0.03:1 (Ni-3Mo-SDC) exhibits the lowest polarization resistance. The cell with that anode and SDC-carbonate composite electrolyte exhibits a maximum power density of 680 mW cm^{-2} at 700 °C, as shown in Fig. 1. The stability of the cell is enhanced with the increase of the content of Mo in the anode.

In order to further improve the cell performance and stability, rare earth elements (La, Pr and Sm) are introduced into Ni-3Mo-SDC. The effects of La, Pr and Sm addition on the catalytic activity for methanol pyrolysis and coking resistance of the Ni-3Mo-SDC anode are examined. The electrochemical performance of the SOFC is enhanced with introducing La and Sm into the anode. The cell with La addition in the anode shows a maximum power density of 765 mW cm^{-2} at 700 °C with methanol as fuel. The addition of La, Pr and Sm into Ni-3Mo-SDC anode improves the stability of the cell, which is mainly attributed to the decreased amount of carbon deposition with a high graphitization degree.

460851 Low Temperature Sintering of Cathode-Supported Micro-Tubular Solid Oxide Fuel Cells Using Sintering Aids

Abstract:

Micro-tubular solid oxide fuel cells (SOFCs) with typical diameters ranging from sub-millimeter to a few millimeters offer several advantages such as facile sealing, high volumetric power density, and better endurance against thermal cycling and thermal shock. Cathode-supported micro-tubular SOFCs, in which a thick cathode layer serves as the structural support, allow for flexible cell-to-cell connections using inexpensive transition metals and also provide better stability under redox conditions. Additionally, cathode-supported cells with a one-end-closed configuration make it possible to use pressurized oxygen as the oxidant, which leads to better power generation performance. However, the fabrication of cathode-supported micro-tubular SOFCs has been found to be challenging because of the high co-sintering temperatures (â‰¥1400 Â°C) required to obtain a dense electrolyte layer on top of the cathode support. At such high co-sintering temperatures, electrically insulating secondary phases are formed due to chemical reactions between cathode and electrolyte materials. Moreover, the high co-sintering temperature gives rise to coarsening of cathode particles and a very low porosity of the cathode support. Thus, it is desirable to reduce the sintering temperature of the electrolyte layer to achieve high-performance cathode-supported SOFCs. In the present work, we fabricated micro-tubular SOFCs with strontium-doped lanthanum manganite (LSM) cathode supports by multi-step dip coating and low temperature co-sintering techniques. NiO and Fe_2O_3 were used as sintering aids to obtain a dense yttria-stabilized zirconia (YSZ) electrolyte at lower co-sintering temperatures. The role of the sintering aids on densification behavior of the electrolyte and the properties of the fabricated micro-tubular SOFCs will be discussed in detail.

458928 Impact of Molecular- and Mesoscales on Macroscopic Thermodynamic and Transport Properties in Perfluorosulfonic-Acid Membrane Using Multiscale Modeling

Abstract:

Performance of polymer-electrolyte-fuel-cells (PEFCs) links strongly to thermodynamic and mass-transport properties of the polymer membrane. The quintessential PEFC-membrane material is perfluorosulfonic-acid (PFSA) polymer, which phase separates into nanoscale hydrophilic water-filled domains through which ions and water transport and into hydrophobic polymer-matrix domains that provide structural integrity and durability. Although considerable progress has been made to reduce reactant mass-transport limitations in PFSA membranes, mass-transport is still a critical area of concern. Aqueous electrolyte transport across a phase-separated membrane is inherently a multiscale problem with aqueous-dissolved species moving through nanoscale domains that are connected to form a transport network at the intermediate mesoscale. To guide the optimization of PEFC material design and operation, we decouple the impact of micro and macro length-scale on mass transport in PEFC membranes.

Specifically, molecular-scale interactions between aqueous-electrolyte and polymer in hydrophilic domains of PFSA membranes are modeled with mean-field local-density theory aligned with an experimentally consistent 3D domain geometry. Our molecular-scale description of membrane conductivity accounts for solvation, electrostatics, solvent dielectric saturation, finite size, and confinement effects. The microscale framework is validated against atomistic simulations and, subsequently, up-scaled to predict macroscale conductivity and solvent absorption by accounting for the interconnectedness of the hydrophilic domains. Excellent agreement is found with experimental data. Our proposed multiscale model for membrane conductivity and solvent sorption provides a new tool to explore potential avenues for improving PFSA membrane performance.

474510 Analysis and Multi-Objective Optimization of Solid Oxide Fuel Cellâ€“Gas Turbine Hybrid Cycle

Abstract:

Abstract:

Chemical process optimization problems often have multiple and conflicting objectives, such as profit, capital cost, operating cost, production cost, energy consumptions, environmental impacts and safety. In such cases, Multi-Objective Optimization (MOO) is useful in finding many optimal solutions, to understand the quantitative trade-offs among the objectives, and also to obtain the optimal values of decision variables. Natural gas, syngas or biogas can be converted into power, heat and electricity, using combustion engine, gas turbine or Solid Oxide Fuel Cell (SOFC). SOFC with Gas Turbine (GT) has shown higher thermodynamic performance, due to better heat integration. Hence, this hybrid conversion system leads to better utilization of natural resource, reduced environmental impacts, and more profit.

SOFC-GT system is divided into five sub-systems: (1) fuel processing using steam reformer, (2) SOFC, (3) anodic GT, (4) cathodic GT, and (5) CO_2 compression. The unconverted fuel from the anodic side of SOFC is combusted in a burner, and then it is used in anodic GT to produce electricity. The SOFC-GT has been simulated in BELSIM-Vali (version 4.7.0.3) flowsheeting software. This study optimizes performance of SOFC-GT system for minimization of both levelized electricity cost and annualized capital cost per kWh, simultaneously. The MOO problem has 12 decision variables, from all five sub-systems of SOFC-GT. The MOO of SOFC-GT system is performed using OSMOSE, which has four integrated parts: (1) stochastic MOO program which provides the values of decision variables, (2) to pass values of decision variables to BELSIM Vali for simulating SOFC-GT system, (3) to obtain temperatures and flow rates of important streams from BELSIM Vali and perform heat integration, and (4) performance evaluation or calculations of objective functions for SOFC-GT system. For optimal SOFC-GT designs, the composite curves for maximum amount of possible heat recovery indicate good performance of the system, and the optimal designs do not require any external hot utility. The SOFC-GT system separate the CO_2, which can be stored and used in other applications. Finally, first law energy and exergy efficiencies of optimal SOFC-GT system are significantly better compared to other conversion systems. The results of this study will be presented at 2016 AIChE Annual Meeting.

Keywords:

Multi-Objective Optimization, Solid Oxide Fuel Cell, Gas Turbine, Heat Integration.

References:

1. Caliandro P., Tock L., Ensinas A.V, Maréchal F., Thermo-economic optimization of a solid oxide fuel cell - gas turbine system fuelled with gasified lignocellulosic biomass. Energy Conversion and Management 2014; 85: 764-773.
2. Facchinetti E., Favrat D., Maréchal F., Design and optimization of an innovative solid oxide fuel cell - gas turbine hybrid cycle for small scale distributed generation. Fuel Cells 2014; 14(4): 595-606.
3. Facchinetti E., Favrat D., Maréchal F., Hybrid cycle SOFC-inverted gas turbine with CO_2 separation. Patent Number - US9228494 B2, 2016.
4. Maréchal F., Palazzi F., Godat J. and Favrat D., Thermo-economic modelling and optimization of fuel cell systems, Fuel Cells 2004; 5(1): 5-24.
5. Palazzi F., Autissier N., Maréchal F., Favrat D., A methodology for thermo-economic modeling and optimization of solid oxide fuel cell systems. Applied Thermal Engineering 2007; 27(16): 2703-2712.

ONSITE

Operation of a Novel SOFC-Battery Integrated Hybrid Generator for Telecommunication Energy Systems

erich erdle

2016 AIChE Annual Meeting
San Francisco - November 13-18, 2016

1

contents

- general project information - project data and consortium
- the energy situation within the telecommunication industry
- a glimpse on trends in the telecommunication industry
- needs of the telecommunication industry
- ONSITE's mission
- ONSITE's achievements so far
  - system design
  - key power sources sofc and NaNiCl-battery
  - first proof of concept test
  - commissioning of sofc generator
- concluding remarks

2

general project information

- overall objective:
  construction and operation of a containerized a hybrid sofc/NaNiCl-battery (SNC) system that generates more than 10 $kW_{electric}$ at high efficiency (>40% electrical, >85% total) and economically competitive costs to prove the concept
- budget: 5.525.540 Euro (3.012.0338 Euro funding from FCH JU)
- duration: 3 years and 9 months
- partners:

Consiglio Nazionale delle Ricerche, Bonfiglioli Vectron, efceco, ERICSSON, FIAMM Energy Storage Solutions, Hes·so, HTceramix, IEN

3

the situation

- telecommunication industry consumes up to 1% of the total electricity of a country
- electricity cost is a dominant cost factor, reduction via more efficient power generation is important
- the users of telecommunication services request 100% availability
- the electric grid is <u>not</u> 100% reliable in many countries
- reliability is key, not only for the customers but also for the operators, blackouts cause big financial losses

Country	Network	Energy Consumption	% of Country Total Energy Consumption
USA	Verizon 2006	8.9 TWh	0.24%
Japan	NTT 2001	6.6 TWh	0.7%
Italy	Telecom Italia 2005	2 TWh	1%
France	France Telecom-Orange 2006	2 TWh	0.4%
Spain	Telefonica 2006	1.42 TWh	0.6%

source: Steve Roy: Energy Logic for Telecommunications; Emerson

AVERAGE COST FOR ONE HOUR OF POWER INTERRUPTION.

Industry	Amount (U.S.)
Cellular communications	$41,000
Telephone ticket sales	72,000
Airline reservation system	90,000
Semiconductor manufacturer	2,000,000
Credit card operation	2,580,000
Brokerage operation	6,480,000

Source: U.S. Department of Energy

4

ONSITE

the illustration of the challenge (1)

- additionally telecommunication companies
  - will have to serve a steadily increasing number of customers in the future
  - and they have to offer many new energy-intense services (e.g. high speed and high volume data communication)
- in 2010 there were 5.3 billion mobile subscribers (that's 77 % of the world population)
- this number has been growing daily
- at the same time, geographically, more areas have to be covered, especially areas where no or only unstable electric grid is available
- thus, there is need for more power and for reliable, grid-independent power

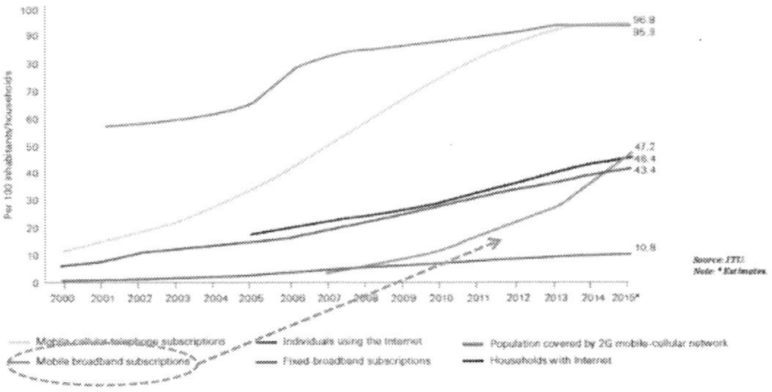

5

ONSITE

the illustration of the challenge (2)

3G mobile-broadband coverage is extending rapidly and into the rural areas

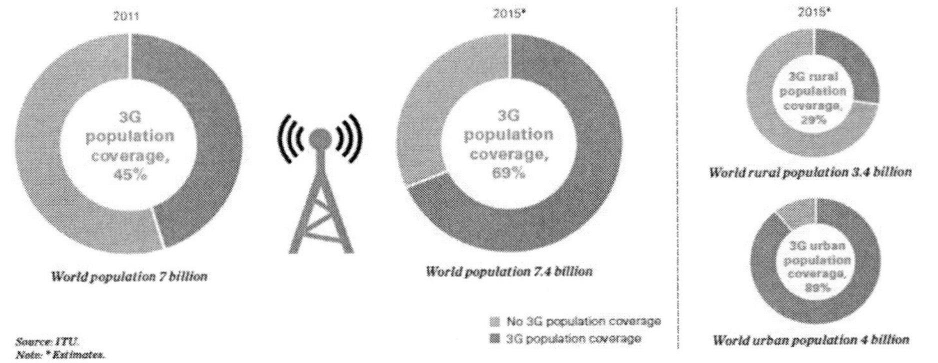

6

needs of telecommunication industry

situation (source: a white paper from EMERSON):

- the cost of and the demand for energy are on the rise, but stable utility power is not always available, especially in remote areas and underdeveloped parts of the world
- despite these challenges, demand for wireless solutions around the world is increasing, prompting service providers to seek and wisely use multiple energy sources
- hybrid energy technologies are emerging as an answer
- there is no "one-size-fits-all" solution. Differences in climate, social environments, site objectives and usage expectations often differ from one site / region to the next, and thus, what works well in one application may be inadequate for another.

needs:

- reduction of environmental impact and CO_2 footprint
- consequent reduction of energy consumption (and OPEX in total), i.e. more efficient systems

the "ONSITE" technical concept aims at

- supply for both on-grid and off-grid Information Technology (IT) plants
- supply for both small Radio Base Stations (RBS) (1-10 kW) and medium sized Data Centers (50 kW +)
- integration with Renewable Energy Systems (RES, e.g. solar PV, wind) minimizing their unpredictability
- integration of the IT plants into the future smart grids/smart cities

7

ONSITE's mission

the "ONSITE" project

- shall accomplish a proof-of-concept of an intelligent supply system (> 10 $kW_{electric}$) for telecommunication equipment and infrastructures (e.g. radio base stations, data centers)
- features a novel hybrid concept that integrates a SOFC (Solid Oxide Fuel Cell) and high temperature Sodium-Nickel-Chloride (SNC) batteries
- aims at
  - better utilization of primary energy
  - higher power generation efficiency
  - use of the waste heat from the fuel cell for heating or cooling purposes
  - significant reduction of the impact on the environment and the climate

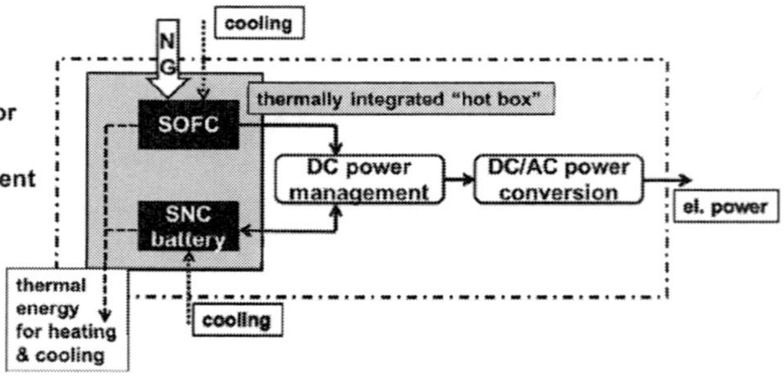

- thanks to its multi fueling capability SOFC can provide off-grid power for telecommunication services in remote areas with no or limited electric and/or gas grid, e.g. by operating on LPG

8

system design - electrical scheme

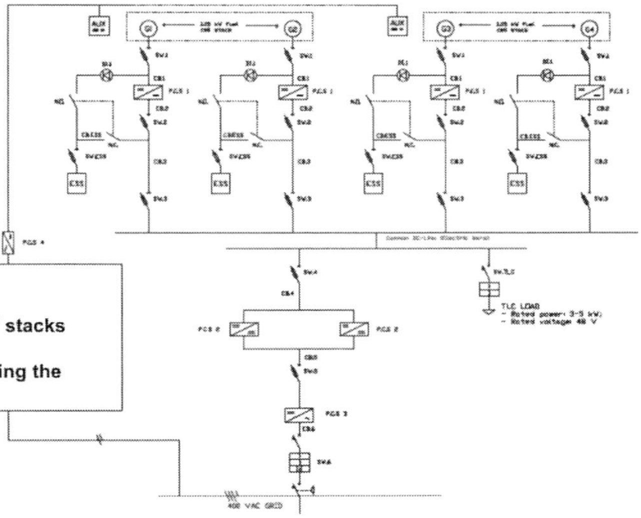

- the system will consist of two subsystems
- each subsystem comprises
  - a 2,5 kW SOFC Hotbox containing 2 1,25 kW stacks and 2 batteries with 3 kW power each
  - each stack has its own DC/DC converter taking the stack voltage to battery level

9

SOFC generator features

feature	values
SOFC stacks operating voltage range	25 to 44 Vdc
fuel	natural gas
nominal power	2.5 kW
SOFC generator operating voltage (regulated)	400 Vdc
operating ambient temperature	0 to +60 °C
SOFC stack operating temperature	750 ÷780 °C
exhaust temperature	< 80°C
temperature of recovered heat	45-65°C

10

SNC battery characteristics

characteristics	values
nominal voltage	48 Vdc
operating voltage range	42 to 56 Vdc
nominal capacity	200 Ah
maximum discharge current	50 A
operating temperature	270 °C
operating ambient temperature	- 20 to +60 °C

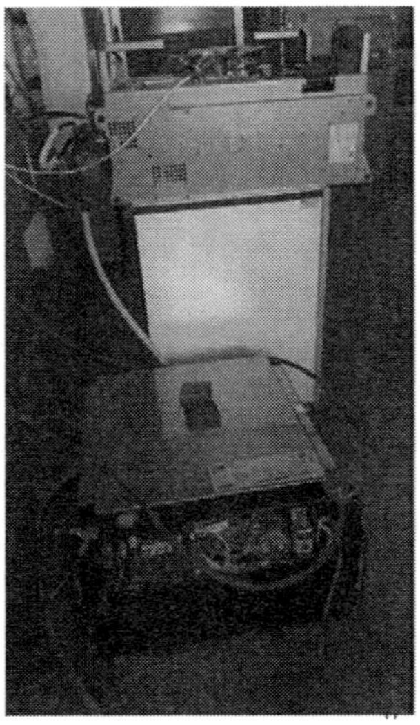

achievements: first proof of concept test (1)

first successful test of SOFC/SNC hybrid generator

SOFC SNC battery

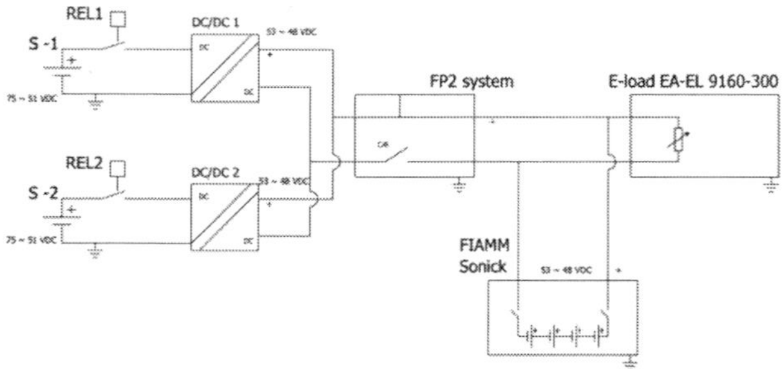

12

achievements: first proof of concept test (2)

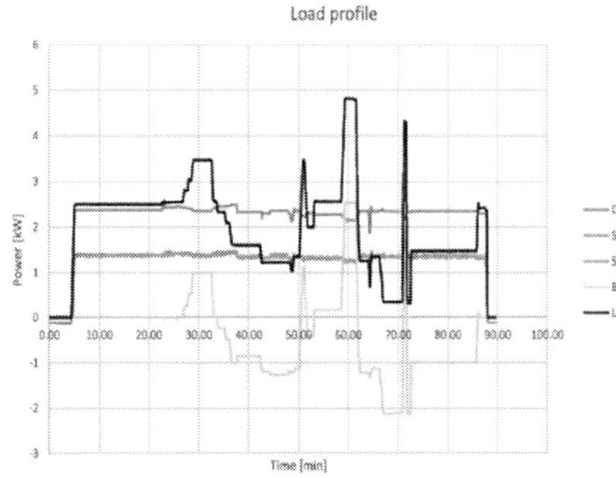

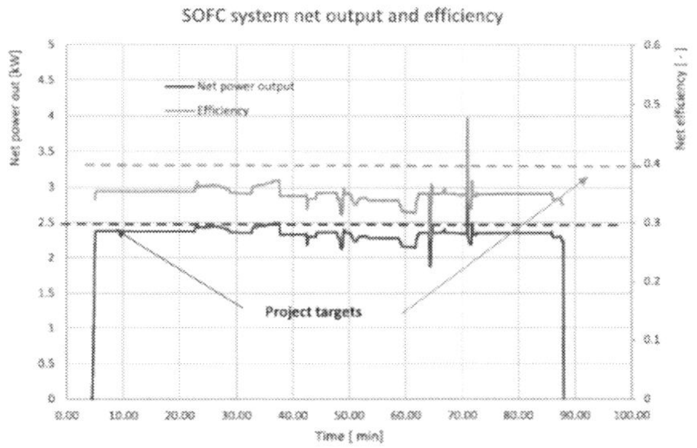

13

commissioning test of sofc system (1)

sofc system running on natural gas

upper compartment: sofc stacks

lower compartment: bop

14

ONSITE

commissioning test of sofc system (2)

480 hours endurance test of sofc system, electrical efficiency approaching 40%, total efficiency above 85%

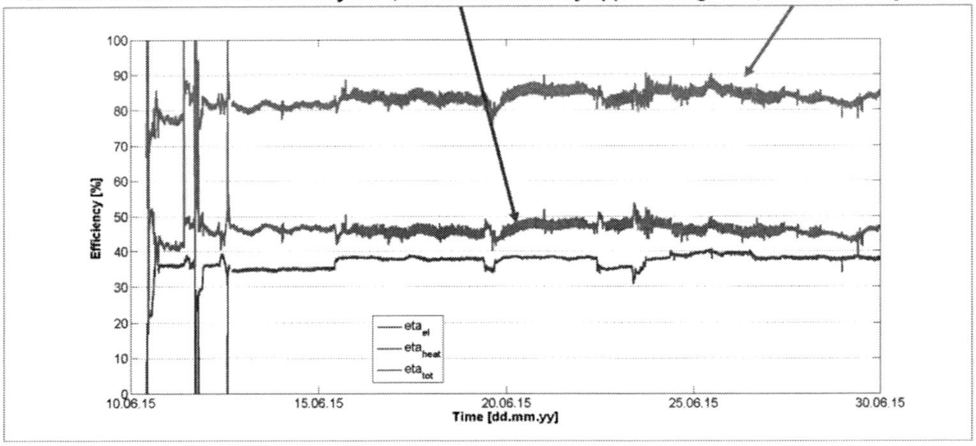

15

ONSITE

commissioning test of sofc system (3)

load variation test of sofc system

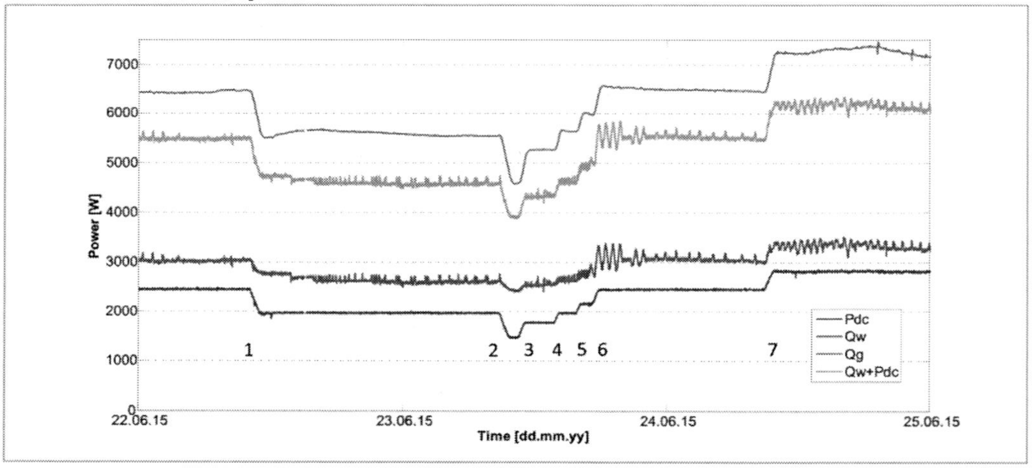

16

commissioning test of sofc system (4)

- electrical efficiency of the stacks and of the system
- parasitic losses and efficiency of DC/DC converter have to be and will be improved

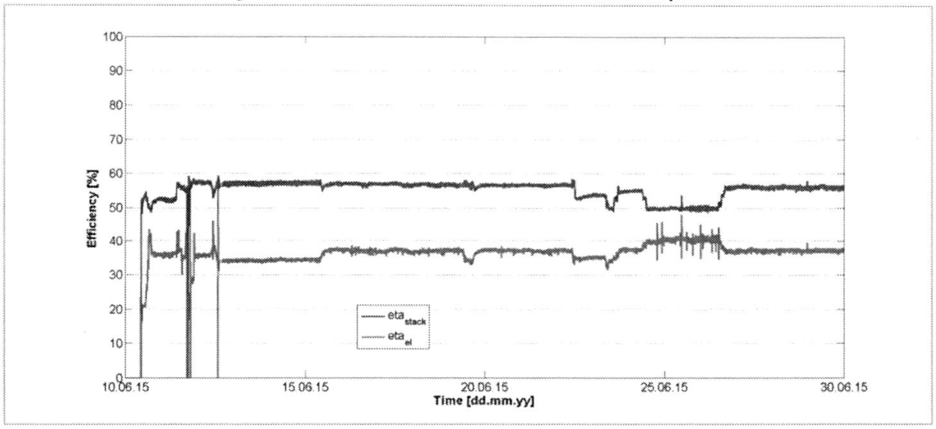

17

commissioning test of sofc system (5)

electric, thermal and total efficiency and electrical and thermal power at different loads

P_{el} [W]	P_{th} [W]	eff_{el}%	eff_{th}%	eff_{tot}%
2000	2700	38%	45%	83%
2500	3000	38%	45%	83%
2800	3400	40%	48%	88%

project goals approached and met at high load, respectively

18

concluding remarks

- power demand in the telecommunication industry is globally growing rapidly
- reliability of power supply has to be improved
- decentralized power generation systems can contribute to solve these issues
- ONSITE is pursuing a new approach by
  - continuous operation instead of backup power concept
  - replacing the battery packs for backup in off-grid locations by a hybrid generator
- due to that and by using a sofc the system can also provide heat and be embedded in smart grids

- Acknowledgements:
  - Funding from the European Union's Seventh Framework Programme for research, technological development and demonstration via its Fuel Cells and Hydrogen Joint Undertaking under grant agreement no. 325325 is gratefully acknowledged
  - I want to thank
    - the entire consortium of ONSITE for the excellent collaboration
    - in particular my co-authors V. Antonucci, G. Brunaccini, M. Ferraro and F. Sergi from CNR ITAE as well as S. F. Au from HTceramix

19

Thanks for listening!

Questions?

www.onsite-project.eu

www.efceco.com

20

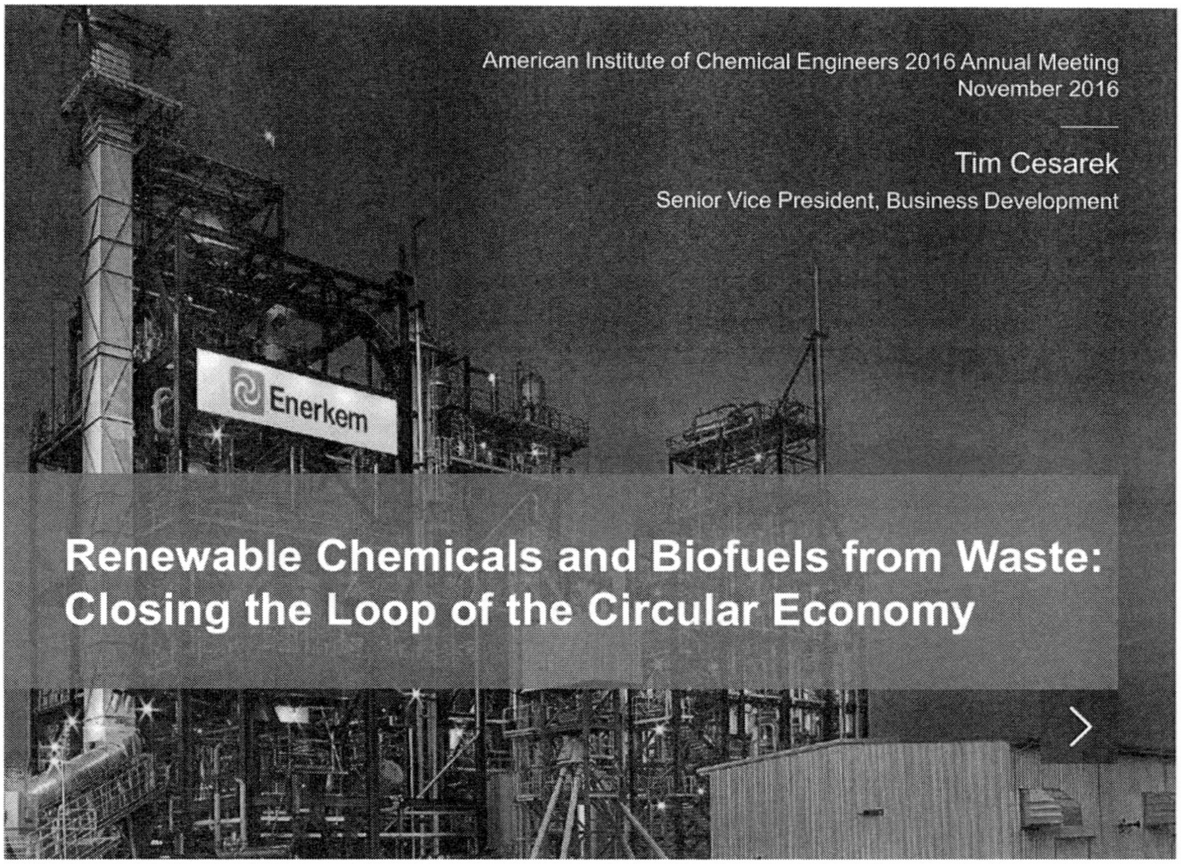

Enerkem at a glance

- MSW-based biofuels and renewable chemicals producer
- World's first full-scale commercial MSW biorefinery beginning operations in Edmonton, Alberta
- $400M invested to date to move from R&D to commercial stage
- 200 employees
- New alternative to landfilling and incineration

We're building the bioeconomy. | © Enerkem, 2016

Enerkem

Methanol sale and delivery

We're building the bioeconomy. | © Enerkem, 2016

Enerkem

Proprietary thermochemical technology and process

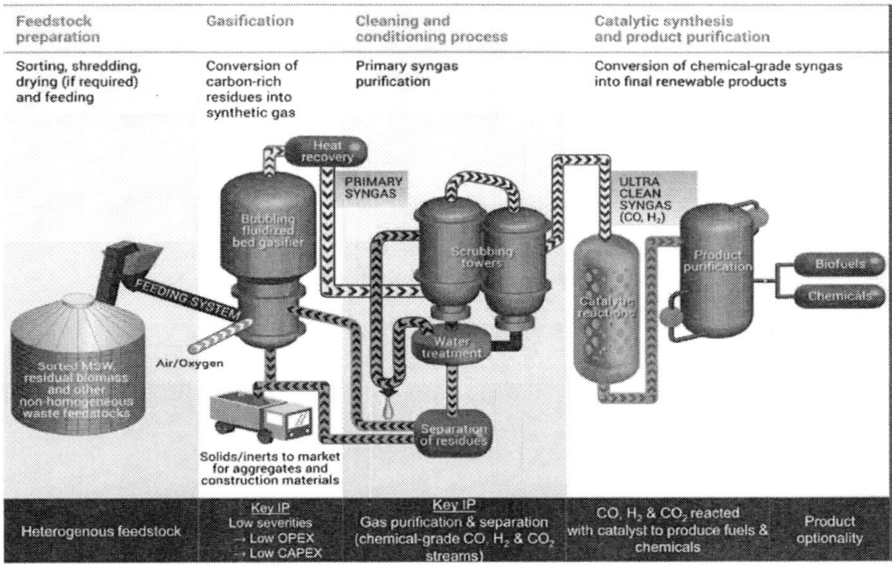

Low-severity technology capable of converting heterogeneous feedstock to clean syngas, the fundamental building block for all Enerkem's products

Key market drivers for waste as feedstock

- Increased scarcity of urban landfill airspace and societal desire for waste diversion
- Circular economy or "cradle-to-cradle" approach
- Low cost unconventional feedstocks
- Renewable fuels mandates around the world
- Consumer pull for renewable and biobased products
- Focus on carbon footprint and GHG emissions reduction

Large market potential

MSW IN THE WORLD

RECYCLED / COMPOSTED

85% TO LANDFILL

1.3 BILLION
METRIC TONS OF MSW
GENERATED PER YEAR

548 MILLION
METRIC TONS OF MSW
SUITABLE FOR ENERKEM'S TECHNOLOGY PLATFORM

THE POTENTIAL:
261 BILLION
METRIC TONS OF METHANOL
USING ENERKEM

Source: World Bank, 2012

We're building the bioeconomy. | © Enerkem, 2016

Enerkem

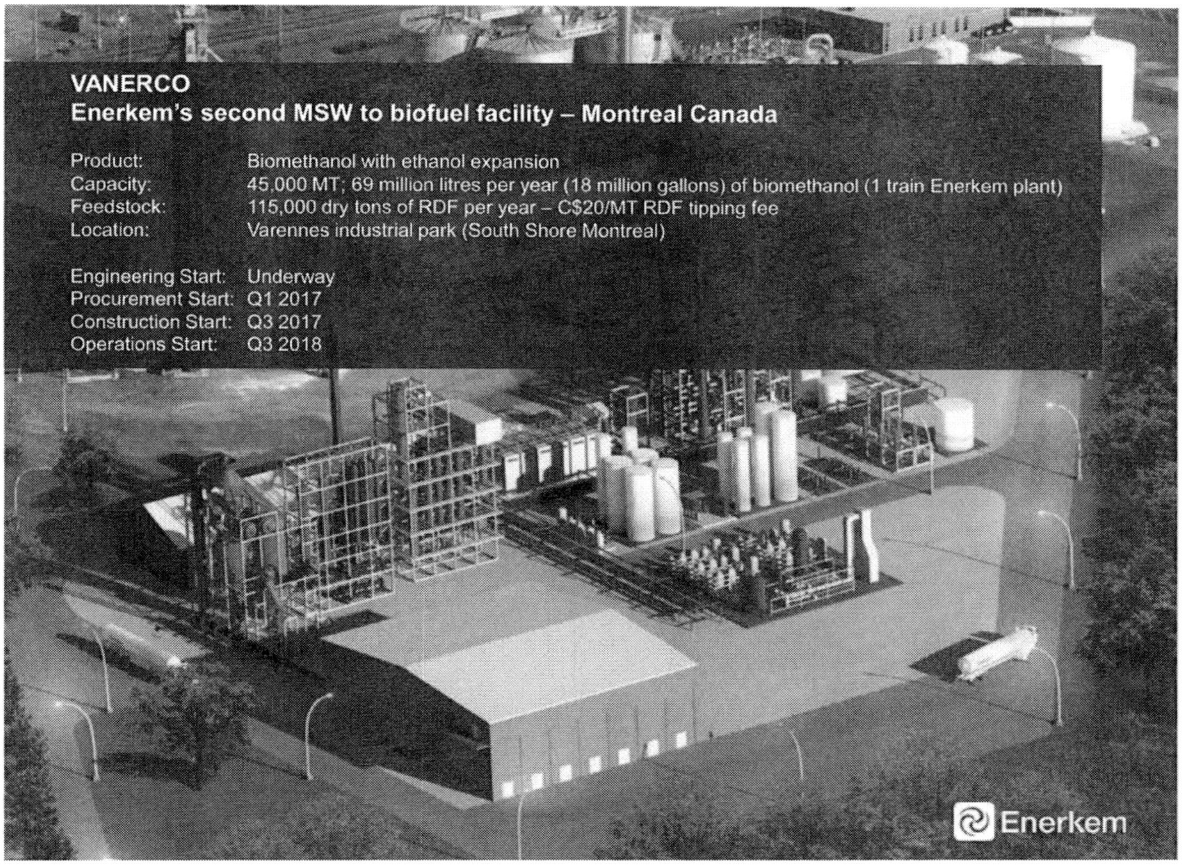
VANERCO
Enerkem's second MSW to biofuel facility – Montreal Canada
Product: Biomethanol with ethanol expansion
Capacity: 45,000 MT; 69 million litres per year (18 million gallons) of biomethanol (1 train Enerkem plant)
Feedstock: 115,000 dry tons of RDF per year – C$20/MT RDF tipping fee
Location: Varennes industrial park (South Shore Montreal)
Engineering Start: Underway
Procurement Start: Q1 2017
Construction Start: Q3 2017
Operations Start: Q3 2018
Enerkem

Biofuels from waste increase energy
diversity and reduce GHG emissions

Trends towards Low Carbon Transportation Fuels

Gasoline, Gasoline Blendstock or Replacement	Definition	Carbon Intensity (including land use effect)
CARBOB	CARBOB	99.78
Corn ethanol*	Derived from corn	75.97
Cellulosic Ethanol*	Derived from corn stover	41.05
Sugarcane Ethanol*	Derived from Brazilian sugarcane	56.66
Electricity	CA grid electricity	105.16
Hydrogen	Compressed H2 from central reforming of NG	151.01

** California Air Resources Board temporary CIs for fuels with indeterminate CIs*

Source: California ARB, July 31, 2015 (Proposed re-adoption of the LCFS: Third 15-Day Modified Regulation Order, Table 6. Tier 2 Lookup Table for Gasoline and Diesel and Fuels that Substitute for Gasoline and Diesel and Table 7. Temporary FPCs for Fuels with Indeterminate CIs)

Using waste as a feedstock for the chemical industry

Waste-to-Chemicals public-private partnership in Europe

AkzoNobel | Enerkem | AIR LIQUIDE

van Gansewinkel | AVR.

Clean Tech Delta | Port of Rotterdam

Innovation Quarter | Gemeente Rotterdam

Provincie Zuid Holland | rotterdam partners

Location of Project in Port of Rotterdam

Product pathways of syngas and methanol

CO + H2 (syngas) and methanol are key building-block molecules of chemical industry

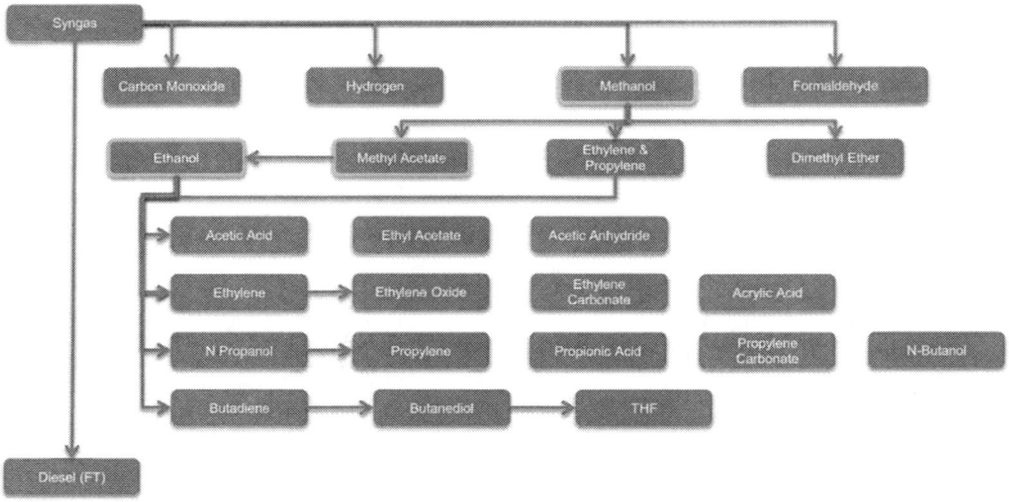

Enerkem

Multi-product platform being developed

Enerkem's platform is being expanded to new products. Acrylic Acid and Acetic Acid are the next steps:

Chemical building blocks in Enerkem's synthesis gas: CO, H_2, CO_2

Product	Global Market Size	Applications
Acrylic Acid	>$14 billion(1)	Industrial Coatings, Paints, Fiber
Acetic Acid	>$4.3 billion(2)	Solvents, Pesticides, and Coatings, Chemical Intermediate
Ethyl Acetate	~$1.5 billion(2)	Coatings, Solvents, Ink
Butanol	>$2.4 billion (3)	Solvent, Chemical Intermediate, Fuel
Propanol	>$3.5 billion (3)	Fuel Additive, Plastic Intermediate, Solvents

Sources:
(1) ICIS, in USD
(2) ICIS, Company estimates
(3) Argus DeWitt

Enerkem

Target growth areas for global partnerships

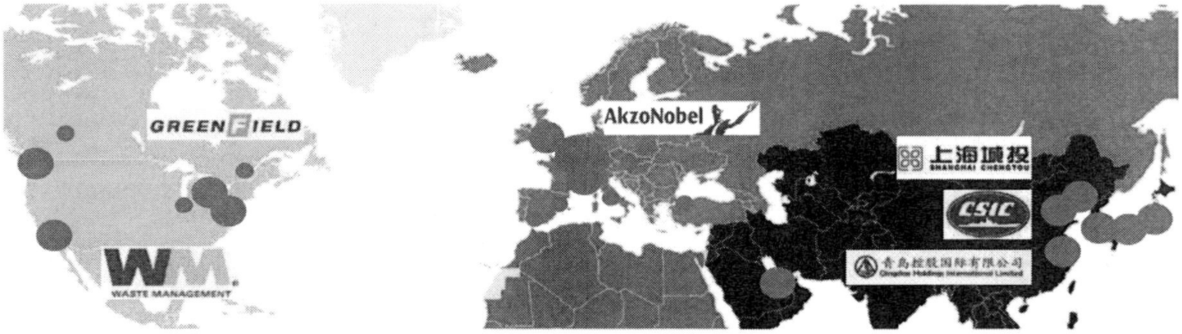

- Strategic partnerships with leading industrial groups
- Selection based on market attractiveness:
  - public policies
  - tipping fees
  - proximity to petrochemical infrastructure
  - population

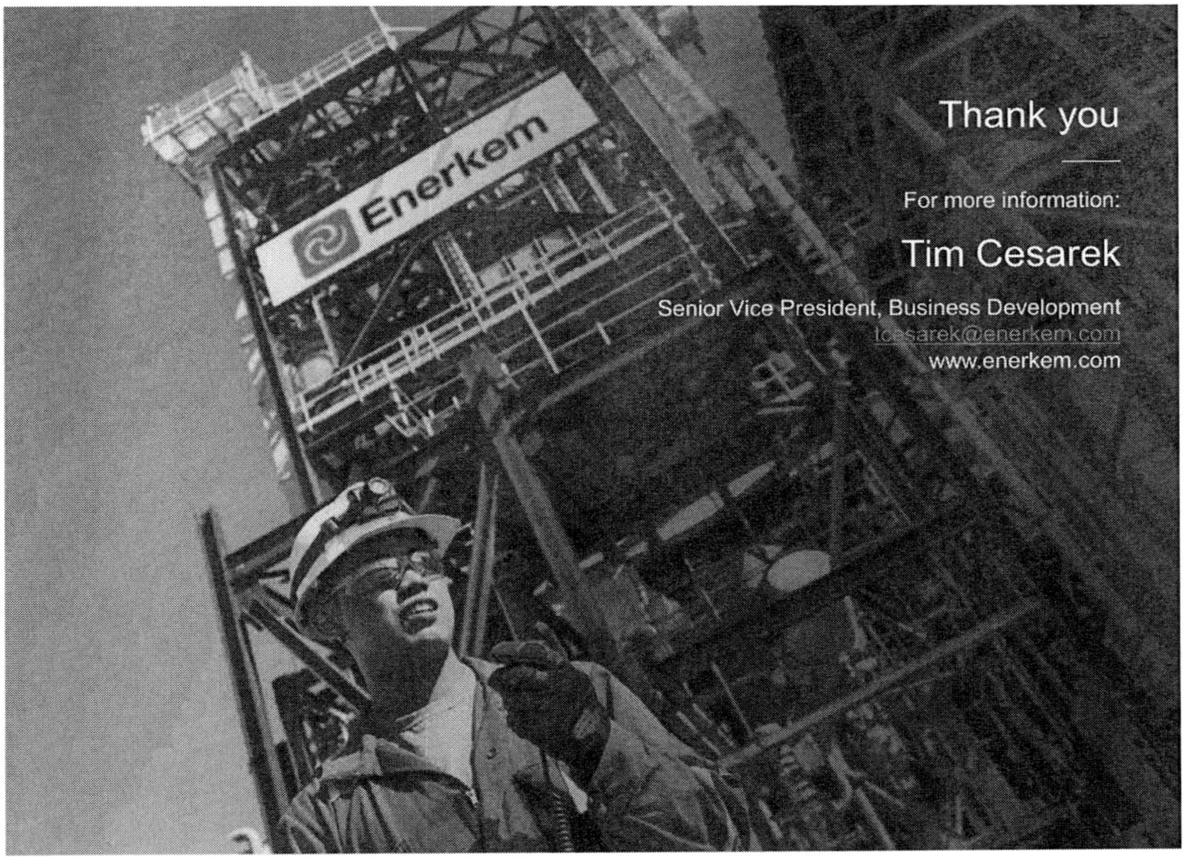

474725 Compact Steam Reformers for Power and Hydrogen Generation

Abstract:

Precision Combustion, Inc. (PCI) is developing high heat flux, compact, thermally-integrated steam reformer technology. The innovation consists of intimately coupled catalytic endothermic and catalytic exothermic sections implemented via novel metal mesh (Microlith$^{Â®}$) supported catalyst. This results in high heat flux for improved product selectivity due to uniform temperature distribution, and flameless operation results in lower peak operating temperatures for increased material safety. These compact reformers are being implemented for fuel cell power generation and H_2 generation. Compared to conventional designs, PCIâ€™s approach offers improved performance, durability and power density. This is facilitated by the use of high surface area, low pressure drop Microlith$^{Â®}$ substrates coated with proprietary catalyst formulations. Additionally, implementation of flameless catalytic oxidation avoids high flame temperatures, increases thermal uniformity, distribution, durability and control. The burner component can utilize system waste heat, anode exhaust or supplemental fuel (as needed) to drive the endothermic steam reformer for efficiency advantages. Performance of the steam reformer has been rigorously evaluated for hundreds of hours with conventional as well as renewable fuels, including at high pressures for enabling hydrogen separation and purification. Data from these tests will be presented and discussed, demonstrating stable, equilibrium limited performance without coking or catalyst degradation. The sulfur tolerance studies of the burner and the steam reforming catalyst will also be presented.

PCI has extensive experience with balance of plant components and controls development. The BOP components e.g. fuel, air and water pumps and controllers are separately packaged, resulting in a compact system. The technology is scalable. The reformer was examined to generate SOFC-quality reformate and was coupled with a water-gas-shift reactor to generate HT-PEM, or PEM-quality (with additional CO cleanup) reformate. Additionally, integration with a hydrogen membrane separator resulted in production of high purity hydrogen. Alternatively, it could also be coupled with a pressure swing adsorber. Data from the reformer as well as from integrated testing with H_2 separator operating on HRJ and S8 will be presented, demonstrating the potential for a compact H_2 generation system. These reformers offer a solution towards compact, fuel flexible power generation systems for various military and industrial applications, as well as for on-site H_2 generation systems.

458513 Development of Smart Ni Based Nano-Oxyhydrides for Hydrogen Production from Bioethanol

Abstract:

Development of smart Ni based nano-oxyhydrides for hydrogen production from bioethanol

L. Jalowiecki-Duhamel[1,*], Y. Romani,[1] W. Fang[1,2], C. Pirez[1], S. Paul[1], H. Jobic[1,3], F. Dumeignil[1]

[1]*Univ. Lille, CNRS, Centrale Lille, ENSCL, Univ. Artois, UMR 8181 - UCCS - Unité de Catalyse et Chimie du Solide, Lille, 59000, France*
E-mail: louise.duhamel@univ-lille1.fr
[2]*Key Laboratory of Medicinal Chemistry for Natural Resource - Ministry of Education, School of Chemical Science and Technology, Yunnan University, Kunming, 650091, China*
[3]*IRCELyon Institut de Recherches sur la Catalyse et l'Environnement de Lyon, 69626 Villeurbanne Cedex, France.*

At the world level, people are concerned by their health and even life degraded by environmental pollution. Therefore, the potential benefits of a hydrogen economy coming from renewable resources are creating a large consensus and increased attention is focused on hydrogen production technologies. So far, hydrogen has been produced from ethanol by steam reforming, partial oxidation, oxidative steam reforming, and autothermal reforming reactions. The endothermic steam reforming reaction extracts more hydrogen atoms from ethanol and water, however, additional energy supply is needed. One alternative way of supplying heat is to add oxygen or air to the feedstock and simultaneously to burn a portion of ethanol. [1-3] Moreover, in complement, some chemical energy can be also brought by a "smart" catalyst, such as an oxyhydride. [4-7]

Well tuning the preparation, formulation and conditions applied, different series of nickel based nano-oxyhydride (Ni-Ce-(Al)-O and Ni-Mg-Al-O) catalysts were developed for the highly efficient and sustainable H_2 formation from ethanol in the presence of oxygen and water with low energy input. The chemical energy released from the reaction between hydride species stored in the nano-materials and O_2, is used to obtain continuous complete conversion of ethanol with simultaneously production of H_2 with an oven temperature at only 50°C (Figs 1 and 2). The influence of different parameters on the activity and selectivity was analyzed, such as the reaction temperature, water-oxygen-ethanol feed compositions, as well as Ni content and in-situ pre-treatment in H_2 of the catalysts. Moreover, different physico-chemical characterizations were performed allowing a discussion on active site and mechanism.

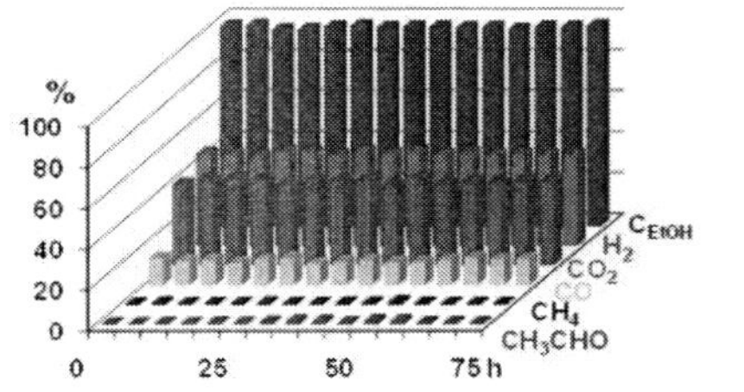

Fig. 1: Ethanol conversion and products distribution on Ni-Ce-O based nano-oxyhydride.

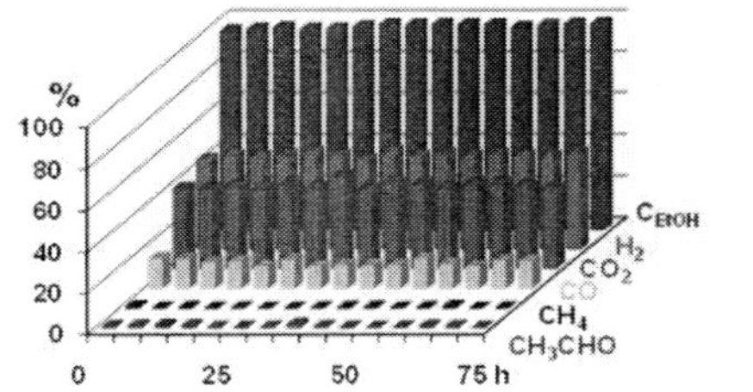

Fig. 2: Ethanol conversion and products distribution on Ni-Mg-Al-O based nano-oxyhydride.

1. G. A. Deluga, J. R. Salge, L. D. Schmidt, X. E. Verykios, *Science* **30**, 5660, 993 (2004).
2. V. Subramani, C. Song, *Catalysis* **20**, 65 (2007).
3. L. V. Mattos, G. Jacobs, B. H. Davis, F. B. Noronha, *Chem. Rev.* **112**, 4094 (2012).
4. W. Fang, C. Pirez, S. Paul, M. Capron, H. Jobic, F. Dumeignil, L. Jalowiecki-Duhamel, *ChemCatChem* **5**, 2207 (2013).
5. C. Pirez, M. Capron, H. Jobic, F. Dumeignil, L. Jalowiecki-Duhamel, *Angew. Chem. Int. Ed* **50**, 10193 (2011).
6. C. Pirez, W. Fang, M. Capron, S. Paul, H. Jobic, F. Dumeignil, L. Jalowiecki-Duhamel, *Appl. Catal. A* **518**, 78 (2016).
7. W. Fang, C. Pirez, S. Paul, M. Jiménez-Ruiz, H. Jobic, F. Dumeignil, L. Jalowiecki-Duhamel, *Int. J. Hydrogen Energy* (2016) DOI: 10.1016/j.ijhydene.2016.04.019.

- Abstract Text

466889 Exploring Transition Metal Carbides and Phosphides for Ex-Situ Catalytic Fast Pyrolysis

Abstract:

Hexagonal Fe_2N type molybdenum carbide (Î²-Mo_2C) has been explored as biomass upgrading catalyst in recent decades for model compounds via hydrodesulfurization (HDS), hydrodenitrogenation (HDN), and hydrodeoxygenation (HDO). Molybdenum carbide has also been shown to be a superior catalyst support, for example, for platinum in comparison to carbon in electrocatalytic water splitting. However, there have not been any studies to investigate the effects of supporting biomass upgrading catalysts such as nickel phosphide (Ni_2P) on molybdenum carbide. Additional carbide-phosphide interfacial sites thus generated have the potential to be of tremendous importance in biomass conversion catalysis. One of the main stumbling blocks for biomass derived pyrolysis oils finding wider use in energy and chemical feedstock generation is the presence of significant amount of oxygen, unlike in fossil reserves. While model compound studies using phosphide and carbides have shown encouraging HDO activities, only a limited number of studies have explored the HDO catalysis on abundant biopolymers such as lignin and cellulose. This presentation will be based on the ex situ catalytic fast pyrolysis of lignin and cellulose in the presence of molybdenum carbide supported nickel phosphide. Particular emphasis will be devoted to the characterization of the interfacial sites and the effect of different Ni_2P loadings on catalysis. Stability of the carbide support during the synthesis of the composite material will also be discussed. In addition, a detailed analysis of product fragments that originate due to the presence of interfacial catalyst sites will be presented via comparison with catalysis results from Mo_2C and Ni_2P individually.

469915 Microwave-Assisted Fast Pyrolysis (MW-FP) of Several Lignocellulosic Feedstocks

Abstract:

Using fast pyrolysis processes, renewable biomass resources can be converted to bio-oils that retain more than 60% of the original energy content, and these oils can then be upgraded to motor fuels or other commodity chemicals. In order to attain economic feasibility, feedstocks for commercial-scale fast pyrolysis must be obtained from a diverse set of low-cost resources for which the conversion performance is well-understood. This presentation summarizes the development of a microwave-assisted fast pyrolysis (MW-FP) reactor that employs a 3 kW microwave beam to heat samples of 1 to 2 grams at rates as fast as 200 °C/s. Liquid, solid and gas yields from as many as 8 samples per day can be measured and analyzed to determine product qualities. Because the individual experiments can be conducted in minutes rather than hours and only small quantities of material are required, MW-FP is advantageous for parametric studies exploring the effects of feedstock material, conversion temperature, and heating rate on fast pyrolysis product yields. Direct measures of product yields show that oil yield increases with volatile matter and lignin contents and decreases with potassium and sodium contents. Contents of total ash, calcium and magnesium did not appear to measurably affect oil yield on a dry, ash-free material basis. Results from the MW-FP reactor have been calibrated to floor-scale fluidized bed fast pyrolysis tests to ensure reliability of the conclusions.

Biomass to Energy – A novel process

By Rakesh Gupta
Lead Process Engineer, Lockheed Martin Energy
Lockheed Martin Corporation
Moorestown, NJ

Background

Biomass is defined as organic material derived from living organisms especially plants. It is an abundant source of renewable energy and chemicals. It is widely available in the United States and worldwide. Its use is being encouraged worldwide as a source of energy, especially in the area of power production. Several government programs have initiated subsidy programs to help develop the power generation from renewable sources. Since biomass is a renewable source, it can be a significant source of fuel for power generation.

Wood is the largest source of biomass. It can come from crop residues, forest thinning, milling residues, agricultural process residues and even municipal solid waste (MSW).

Biomass constituents include organics such as cellulose and lignin. They also contain water and other components such as chlorine, nitrogen, phosphorus, sulfur, silicon and trace metals.

The most important characteristic of a biomass is its heating value. This value is reduced by the presence of water and ash. Although, water presence in the biomass can also be beneficial by assisting in the steam reformation process discussed later.

Lockheed Martin is committed to developing renewable energy sources and has committed significant resources towards its development. Lockheed Martin has partnered with Concord Blue, licensor of a novel advanced gasification technology to convert biomass to energy. The purpose of this partnership is to develop a scalable and fully integrated solution. The Concord Blue process utilizes a patented method of "staged" reforming to convert organic material into a hydrogen rich syngas. This syngas is cleaned of contaminants and it can be either separated into valuable chemicals or fed to a combustion engine or turbine for power generation.

These power plants can range from 250 kW up to 20 MW in size.

Technology

There are many ways to use biomass as a source of energy for power generation. The most common method is to use it as a co-fuel with other sources such as coal, natural gas, etc. However, biomass can also be gasified using air or oxygen or pyrolyzed using a heat source. Some systems have a combination of pyrolysis and combustion/oxidation processes.

Syngas, short for synthetic gas, is the generation of gas from an organic fuel such as MSW, or woodchips by pyrolysis. In this process, the organic waste is gasified by addition of heat. The gases produced have a higher hydrogen to carbon ratio than the source material. The gases also contain carbon monoxide, hydrogen, methane, tars and other contaminants.

Pyrolysis involves the decomposition of organic compounds into gases and solids. This is done in the absence of oxygen. When the biomass is initially fed to the pyrolyzer, the volatile components are released almost immediately. The remaining char is converted to H_2 and CO in slower reactions.

Steam reformation process involves converting the volatile components such as methane and other organics into hydrogen with endothermic and reversible reactions. This process requires addition of steam and energy. These reactions can be repeated for higher hydrocarbon chains also, although each of those reactions have their own energy requirements.

In the Concord Blue staged steam reformation process, syngas is generated with components such as hydrogen, methane, carbon monoxide and carbon dioxide along with other tars and contaminants.

The raw syngas is cooled and cleaned in a scrubber system and then fed to a combustion engine to generate power, but ultimately the syngas could have other uses as well. The components of the syngas could be separated into valuable chemicals or converted into biofuels.

The process generates many valuable byproducts such as char that may be used for agricultural applications, steam for plant use and heat that could be used for applications such as hot water for district heating. Biochar produced from the pyrolysis process can also be used as a source of fuel for the pyrolysis process. It can also be used as activated carbon feedstock.

Syngas generated from the conversion process can also be converted into synthetic crude oil which can be further refined into clean synthetic fuels and chemicals such as kerosene, diesel, jet fuel, lubricants, etc. It can also be blended into crude oil and processed in a refinery. The practice involves using the Fischer-Tropsch process.

Suppliers

Gasifiers are provided by many suppliers such as GE, Conoco, Shell, Siemens, MHI and KBR. Shell and Siemens systems are oxygen blown using slagging type burner. MHI and KBR technologies use air or oxygen, do not use a burner and are non-slagging type.

The technology licensed by Lockheed Martin from Concord Blue utilizes staged reforming as a method of generating synthetic gas. The organic material is initially pyrolyzed and then reformed using steam and heat.

Pilot Plant

In order to demonstrate this technology, Concord Blue's patented method of "staged reforming", Lockheed Martin has built a pilot scale project at its Owego, NY facility. Lockheed Martin selected wood chips as the initial fuel source, since they are easily obtained in consistent size, they are available in abundant supply and they provide a consistent source of organics. Literature search showed wood chip samples from New York state have heating values of 5,600 Btu/lb. to 7,800 Btu/lb., moisture content of 28 – 42 % and ash content from 0.4 to 1%.

The size of this of this project is 250 kWe. It has the capability of producing renewable / sustainable power using various solid wastes, organic wastes, non-organic residues and woody biomass feedstocks such as clean wood chips and refuse derived fuel (RDF), while showcasing the technology. The feedstock will initially be woody biomass with an average moisture content of 37%.

The following is a simple sketch of the Concord Blue technology.

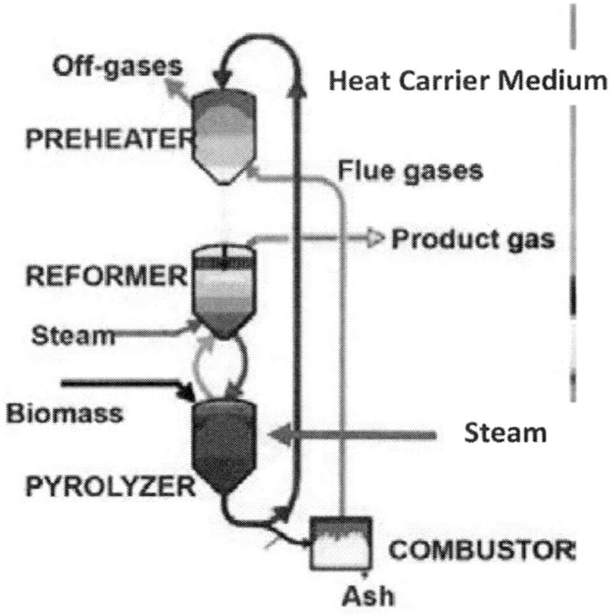

Figure 1 – Concord Blue Staged Reforming Technology
(Institute of Wood Technology and Wood Biology (HTB), Hamburg, Germany)

The biomass is fed to the pyrolyzer at temperatures above 900 °F. The gases generated are reformed in a steam reformer to increase the hydrogen concentration in the syngas. Heat is supplied to the reformer and the pyrolyzer by a heat carrier medium. The heat carrier medium is heated in the preheater as shown above in Figure 1.

The 250 kW Technical Assessment and Demonstration Facility's subsystems include:

- Feedstock Storage and Handling - The biomass is received in specially equipped trucks. It is transferred to the pyrolyzer using a number of conveyors. All feedstock materials are processed off-site by a feedstock provider and transported to the plant. Feedstock processing includes sorting, shredding to size, and removing all ferrous and non-ferrous metals as well as all inert material such as rocks, stones, ceramics, etc.

- Staged Reforming Equipment - The primary function of the Staged Reforming System is to gasify the incoming feedstock to generate syngas for downstream combustion in the Internal Combustion Engine. Biomass is fed to the Pyrolyzer where it gasified in a slow and controlled manner. Gas from the Pyrolyzer is fed to the Reformer where higher temperature reactions take place in the absence of oxygen. It is in this high temperature reformer where the water gas shift reaction occurs. It consumes water and energy to increase the heating value of the syngas by generating hydrogen. This is also where tars are cracked with water to produce hydrogen and carbon monoxide. This step improves the quality of the syngas by reducing the amount of tars that end up downstream into the syngas cleanup equipment potentially fouling piping and the gas cleaning system, thereby reducing gas flow to the generator

- Burner - The Burner is a vendor supplied package, that is designed to combust natural gas, syngas and other fuel sources, such as spent biodiesel, in order to generate hot flue gas to supply heat to Staged Reforming System's Pre-Heater. Biodiesel is atomized into the burner using compressed air. Fresh outside air, using combustion air blower provides the oxygen source for combustion within the burner. During start-up, natural gas is used to fuel the burner and bring the flue gas up to operating temperature. Note that in other installations, the char generated by the process will be combusted to provide the heat for the Staged Reforming System.

- Syngas Cleaning - Raw syngas from the Staged Reforming System Reformer contains a number of contaminants not suitable for running an IC engine, such as acid gases (i.e. H2S, HCl, HF, COS), ammonia, particulate matter, and tars. (The presence and concentration of contaminants is affected by the type of fuel used.) The function of Syngas Processing and Conditioning system is to clean and condition the raw syngas from the Staged Reforming Process so that it conforms to the syngas quality requirements of the internal combustion (IC) engine. The first step in the syngas scrubbing process is the quenching of the hot gases. The gases then pass through a venturi scrubber to remove particulates. The system includes a scrubber to remove ammonia and another scrubber to remove H2S. The syngas pressure is boosted to meet the requirements of the engine. Also, the gases are reheated to reduce the humidity of the gases.

- Flue Gas Cleaning will be added in the future depending upon initial results of environmental stack testing.

- Syngas Storage / IC Engine-Generator - The function of the Power Generation and Gas Storage system is to convert the cleaned syngas into electrical energy through the IC Engine. The gas storage system provides a mechanism to even out the composition of the syngas and provides a few minutes of storage to allow for graceful shutdown of the engine, when required. A single 250 kWe internal combustion engine vendor package is used to convert the cleaned syngas into electrical energy.

- Flare – The flare is used to burn off any off-specification syngas or cleaned syngas in case the engine is not operating. The flare provides safe combustion of the syngas to prevent harmful toxins and combustible materials from emitting to the atmosphere. The flare is designed to maintain a constant flame at the flame tip. Nitrogen is used as the purge gas for the flare velocity seal.

- Wastewater Treatment – The wastewater treatment system collects the water discharge from the scrubbing system. Currently, it is shipped offsite for treatment. In other applications, it would be treated on-site and returned to the plant for process water uses.

- Cooling Water System – The quenched gases in the syngas scrubbing system require cooling for the engine to combust them. They are cooled using a closed circuit cooling tower.

- Compressed and Instrument Air System – The compressed air system provide the air required for the operation of the instruments, valves and the atomizing air for the burner.

- Natural Gas – Natural gas is currently used as the fuel source for heating the heat carrier medium.

- Nitrogen – Nitrogen is used in the process to purge out the system before generating syngas. It is also used in a number of components to prevent mixing of syngas with air.

- Biodiesel – Biodiesel is used as the absorbent for the tars generated in the process. The spent biodiesel is used as a fuel in the burner.

A simplified flow diagram of the system is as follows:

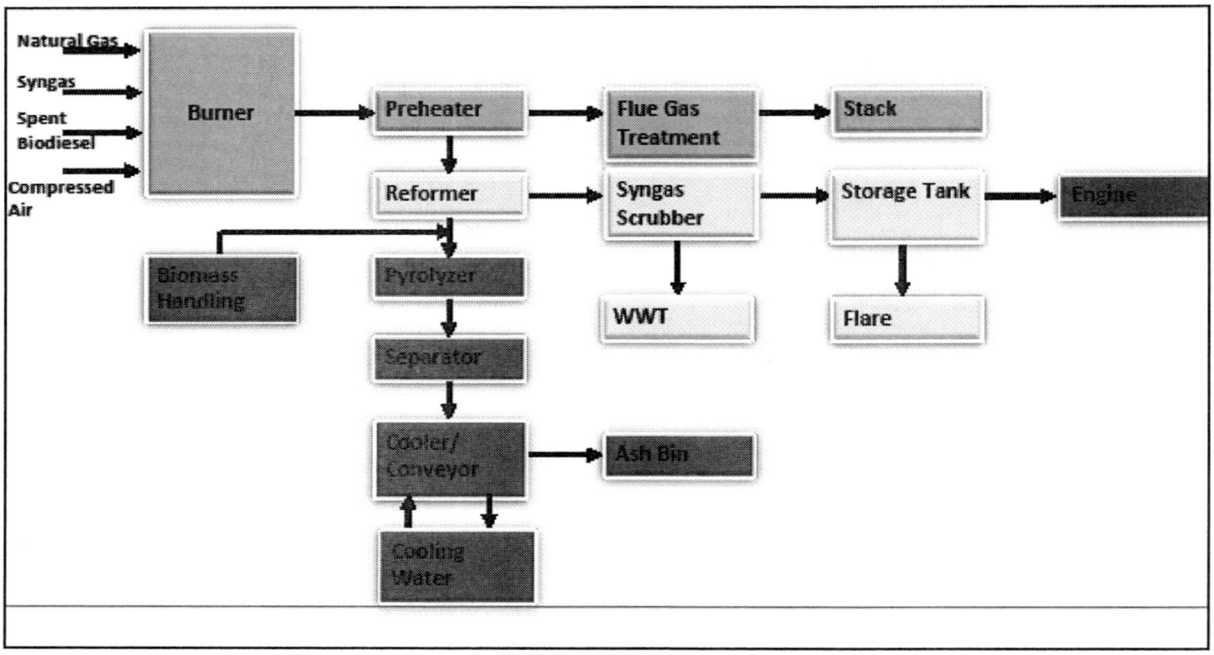

Figure 2 - Block Flow Diagram of the Owego pilot plant

The following is conceptual design of the pilot plant:

Figure 3 – Initial concept of the pilot plant

The following Figure 4 is a picture of the installed system.

Figure 4 – Owego pilot plant

Commercial Scale Plant –Herten, Germany

This is a full scale plant that will be built to generate more than 5 MW of power using wood chips as a fuel. It is similar to the pilot plant described above, except fuel drying has been added to the process.

Other energy-saving strategies are being implemented in this design, including use of biochar as a heat source for the Staged Reforming System, and use of heat from engine exhaust as a heat source for the biomass dryer.

The pictures below Figures 5 & 6 show a conceptual view of the original site plan.

Figure 5 – Herten Biomass Plant Conceptual View

Figure 6 – Herten original site plan

Results

The Owego pilot plant has completed the subsystems commissioning. System level commissioning underway. In the last month, we have made syngas that meets the quality requirements of the engine and have run the engine briefly. Plans are under way to complete the tuning and optimization of the process. Complete results of the tests will be available later this year.

The Herten, Germany project is still under design and will begin construction later this fall.

References

1. "Elemental Analysis of Wood Fuels", Final Report, June 2013, NYSERDA Report 13-13

GENERATION OF HYDROGEN FROM SOLID FEEDSTOCK USING THE HEATPIPE REFORMER TECHNOLOGY WITH IN-SITU HYDROGEN SEPARATION BY NICKEL MEMBRANES

Jonas M. Leimert, Jürgen Karl
Institute of Energy Process Engineering, Friedrich, Alexander-University Erlangen-Nuremberg, Germany

Introduction

The Heatpipe Reformer was developed as an allothermal gasification system at TU Munich during the EU project ``BioHPR" (Ref. ENK5-CT-2000-00311). The concept was commercialized by Agnion Inc., which built a demonstration plant in Pfaffenhofen, Germany in the scale of 500 kW and two commercial plants in the 1 MW scale in Grassau, Germany and Auer, Italy. Results from these plants are given by Gallmetzer in [1].

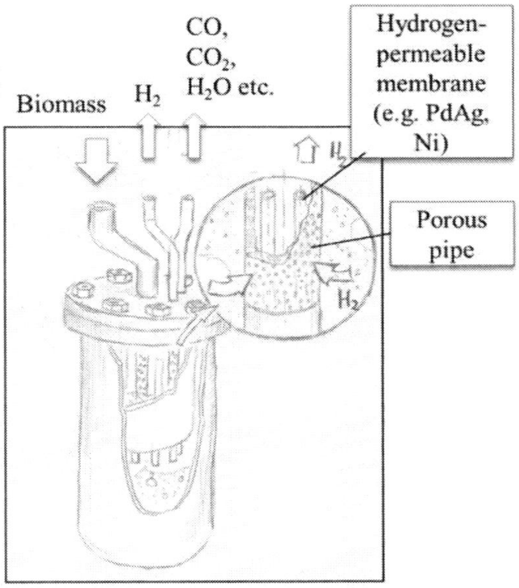

Figure 1: Sketch of Heatpipe Reformer with integrated hydrogen separation by membranes

Figure 1 shows a process scheme of the Heatpipe Reformer: In contrast to other dual fluidized bed gasification processes, the gasification takes place in a pressurized reformer chamber at 2-10 bar and 800°C. The gasification system can process both biomass and coal as feedstock. The pressurized synthesis gas allows usage for SNG synthesis, combustion in engines or gas turbines. Alternatively, the hydrogen in the synthesis gas can be separated for coupled heat and hydrogen generation. This could be achieved by a conventional process chain using a downstream CO shift reactor and hydrogen purification systems. As proposed by Karellas et al. and Schäfer the hydrogen could also be separated directly in the reformer chamber [2], [3]. This system enhances hydrogen generation due to the shift of the water-gas reaction to the product side caused by hydrogen removal:

$$CO + H_2O \rightleftharpoons CO_2 + H_2$$

For this application dense hydrogen permeable membranes with a solution-diffusion mechanism can be used, providing hydrogen with highest purity. Such dense hydrogen permeable membranes have been covered extensively over the years. In the case of palladium membranes the temperature regime of these investigations was in the range of 200-600°C restricted by the kinetics of hydrogen permeation at lower temperatures and the poor durability of Pd/Ag composite membranes at higher temperatures. Also membrane systems with other materials like Niobium, Tantalum, Titanium, Vanadium or Nickel have been proposed with varying limitations regarding temperature, pressure and atmospheres. For the Heatpipe Reformer at FAU-EVT the membrane system has to withstand temperatures of 750-900°C and pressures of up to 5 bar.

State-of-the-art

The approach at the Institute of Energy Process Engineering is the usage of dense nickel capillaries for hydrogen separation due to their low cost and high durability at the desired process temperature. Nickel membranes were already used in the Heatpipe Reformer system at FAU-EVT for hydrogen degassing of liquid metal heat pipes as described in [4]. In this application the nickel capillaries were operated for 2500 h at 800 – 870°C in sodium atmosphere without failure.

This section introduces the mathematical background of hydrogen permeation and the analysis of turnover rates and hydrogen yields. It also shows the state-of-the-art of hydrogen permeable membranes for high-temperature applications.

Hydrogen permeation and process analysis

The permeation of diatomic gases like hydrogen through a dense body is divided into several steps: The adsorption of the hydrogen on the surface, the dissociation into protons and dissolution into the metal, which is sometimes referred to as solution of the hydrogen. The hydrogen diffuses through the metal as protons, which is in most cases the rate-controlling step of the whole process, finally recombines and desorbs from the solid [3], [5], [6].

The following equation describes the hydrogen molar flow through a dense membrane n_{H2}, it depends on the permeability P of the membrane material, membrane area A, thickness Δx and the square root of the pressure difference between retentate and permeate p_1, p_2 respectively.

$$\dot{n}_{H_2} = -P \cdot \frac{A}{\Delta x} \cdot (p_1{}^{0.5} - p_2{}^{0.5}) \quad (1)$$

The permeability itself is a temperature activated material property and can be calculated with the Arrhenius equation using the permeability constant P_0 permeability activation energy E_A and gas constant R at the respective temperature T:

$$P = P_0 \cdot e^{-\frac{E_A}{R \cdot T}} \quad (2)$$

To describe the hydrogen separation in relation to the amount of treated gas, the permeate flux is divided by the maximum amount of hydrogen, which is possible to generate from the feed. The result is the dimensionless hydrogen separation rate Y_M.

$$Y_M = \frac{\dot{n}_M}{4\,\dot{n}_{CH4} + \dot{n}_{CO} + \dot{n}_{H2}} \quad (3)$$

The feedstock in gas phase reaction consists mainly of CO and CH_4. To describe the course of the reaction, the CO and CH_4 turnover X_{CH4} / X_{CO} is introduced:

$$X_i = \frac{\dot{n}_{i,out}}{\dot{n}_{i,feed}} \quad (4)$$

State-of-the-art of high temperature hydrogen permeable membranes

Figure 2 shows a summary of the literature study of high temperature membrane systems. Additionally to the permeability values of Pd based membranes mentioned above, also values from Akamatsu et al. [7], Hatlevik et al. [8], Yan et al. [9], Kusakabe et al. [10], Fernandez et al. [11], Guo et al. [12], Mardilovich [13] and Xomeritakis and Lin [14] are included. Palladium composite membranes show high hydrogen fluxes nearly reaching the physical boundary, which is shown by the hydrogen flux calculated from pure palladium. However, many membranes show fluxes well below this ideal value covering up to three orders of magnitude. The temperature limit for these membranes lies around 650°C making an application in gasification processes very challenging. Further problems of Pd-composite membranes in synthesis gas applications are coke formation, which can lead to membrane failure as carbon atoms diffuse into the Pd lattice leading to cracking [15]. They also show a degradation in hydrogen flow when exposed to synthesis gas components like CO, H_2O or sulfur components like H_2S [15]–[18].

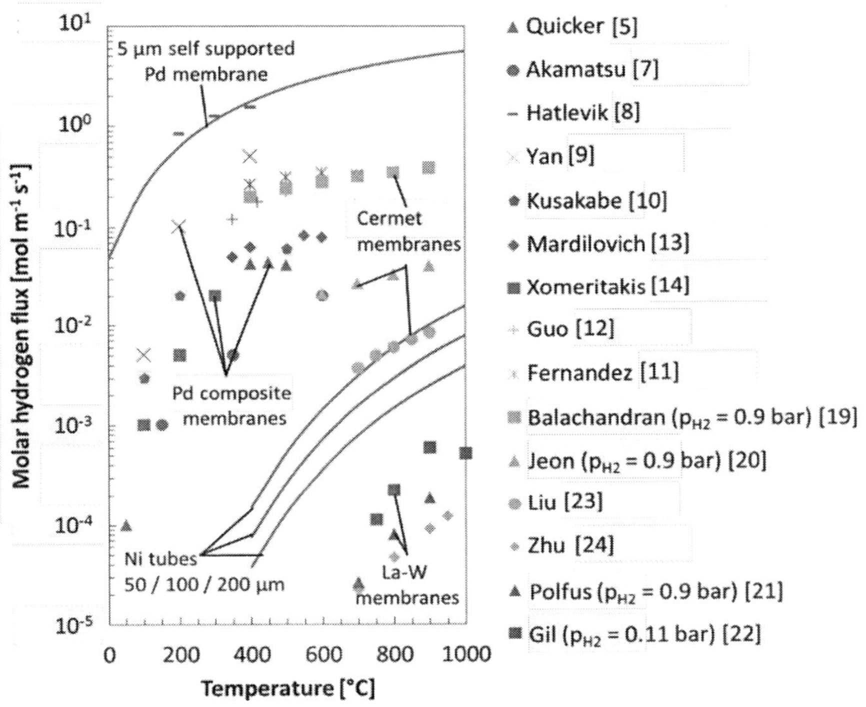

Figure 2: Literature values of membranes for high temperature hydrogen separation calculated from published permeability values at P_{H2} = 1 bar unless otherwise stated

Future membrane solutions could come from cermet or La-W membranes as described in [19]–[24], which combine high molar fluxes with temperature stability and tolerance for many gas components. However, the transfer from laboratory to an industrial application could again pose problems. Recent literature describes only small, disk shaped membranes.

Nickel membranes show in the temperature regime of gasification hydrogen fluxes in the same order of magnitude as some of the Pd and cermet membranes while showing high stability. Considering low nickel prices, this material could be easily used in process engineering without an elaborate membrane fabrication process. When small tube diameters are used, high membrane areas can be accomplished easily by constructing membrane bundles.

Gasification experiments

Fig. 3 shows a CAD sketch of the 100 kW Heatpipe Reformer at EVT-FAU. The reformer pressure vessel is integrated into the combustion chamber, described in a previous publication [25]. Eight heat pipes, filled with sodium, transfer the heat, which is needed for gasification, from the combustor into the reformer. The fluidized bed furnace is made of refractory concrete to withstand the temperatures and abrasive conditions of the fluidized bed. The primary air inlet is located in the lower part of the furnace. Before entering the combustion chamber the primary air is preheated. Therefore, the air is lead through an integrated double walled heat exchanger with a pillow-like structure located at the external wall of the freeboard. The secondary air is provided by four steel pipes with a nominal diameter of DN25 and is blown into the freeboard radially. The flue gas is led out of the combustor at the top with a temperature of about 200-400°C after preheating primary and secondary air.

The reformer is filled with olivine with a particle size of 0.063 - 0.180 mm from Sibelco Europe with a bed height of 1.1 m. Pressurized steam and lignite is fed into the gasifier through a stand pipe. This ensures a thorough mixing of steam, lignite and olivine. The synthesis gas leaves the gasifier with a temperature of about 500-700°C through an upper outlet.

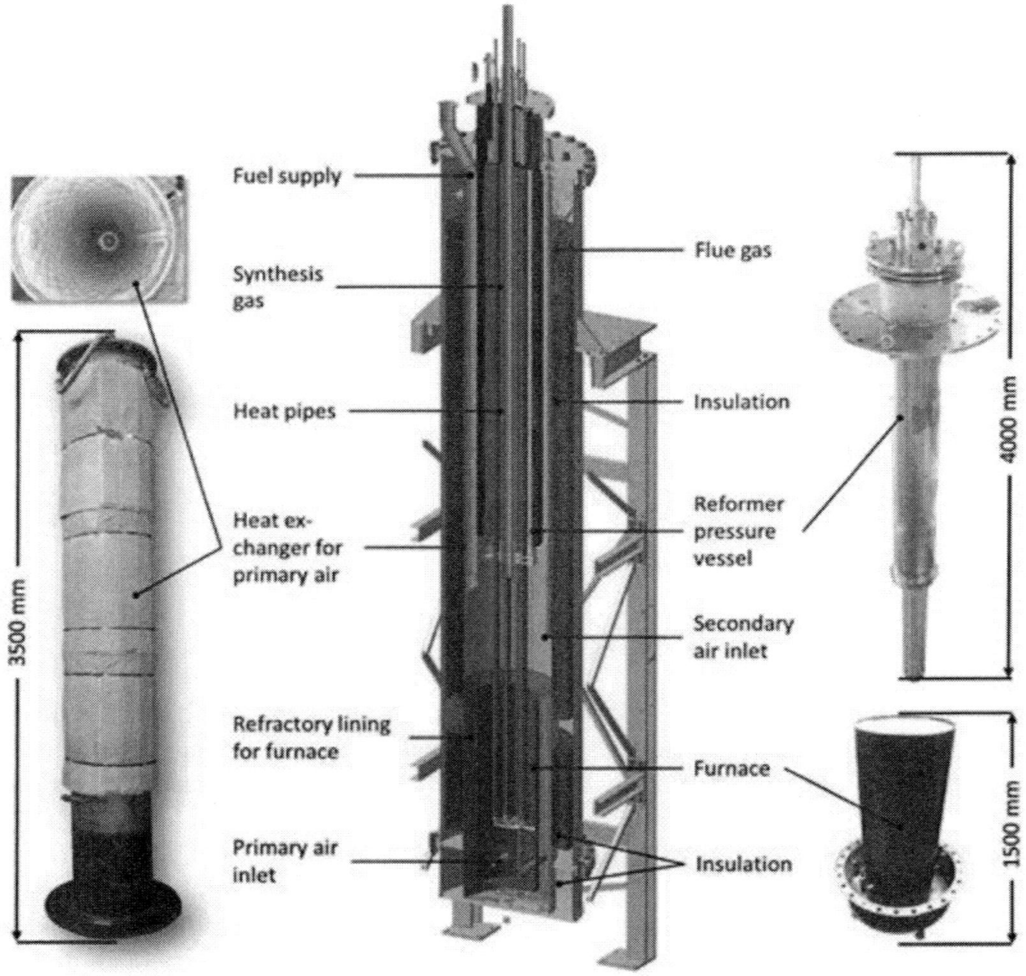

Figure 3: CAD sketch and photographies of the EVT Heatpipe Reformer

Experimental setup

The Heatpipe Reformer described above is integrated in a process chain which guarantees the fuel, air and steam supply as well as the conditioning and analysis of the synthesis gas. The piping and instrumentation diagram of this process chain is shown in Figure 4.

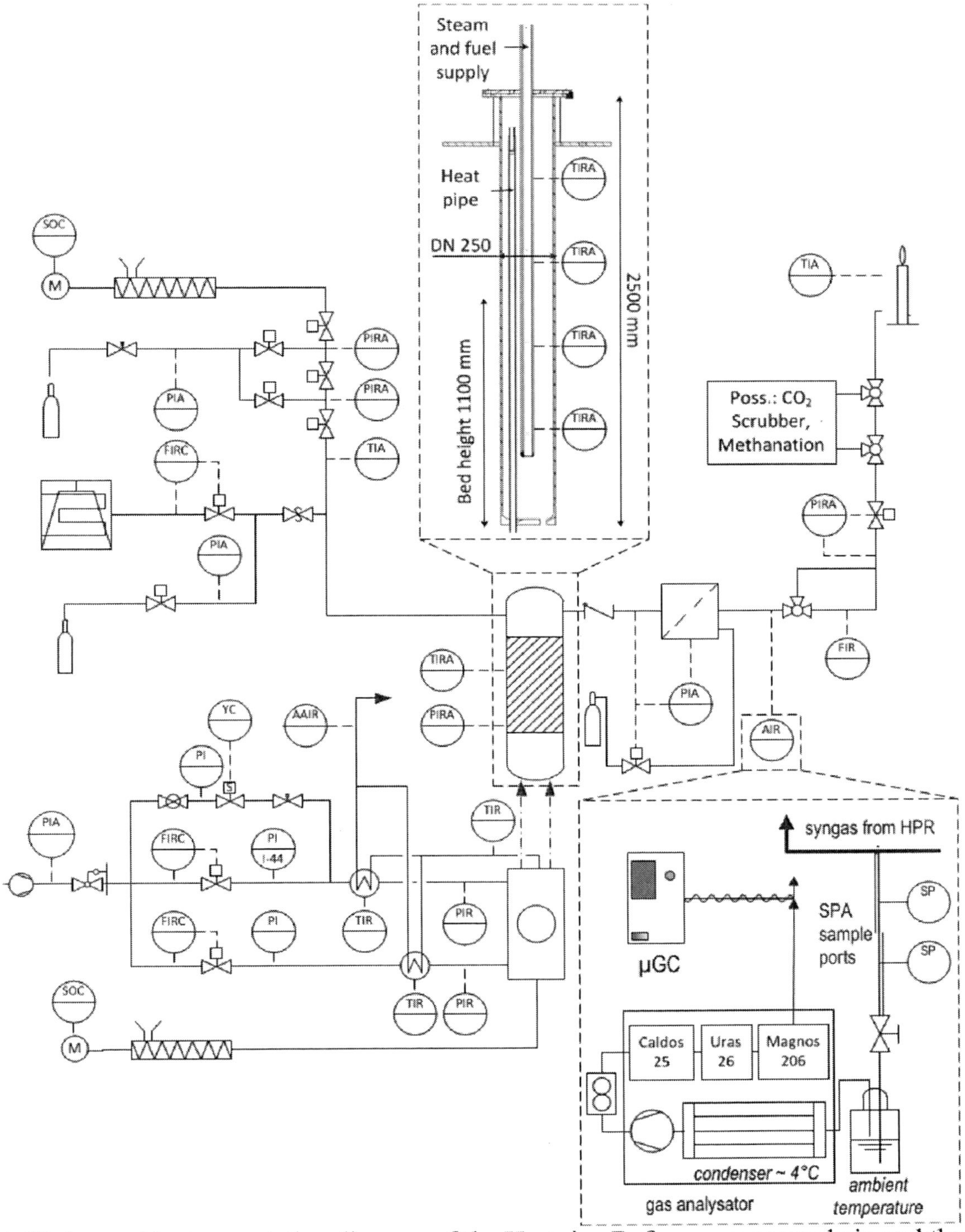

Figure 4: Piping and instrumentation diagram of the Heatpipe Reformer process chain and the gas and tar analysis system

For the combustion, the fuel input is ensured by a screw conveyor. The inlet of primary and secondary air into the combustion chamber is regulated with mass flow controllers fed by compressed air. The lignite supply to the reformer also takes place by a screw conveyor. To pressurize the lignite a lock hopper system with three ball valves is installed above the reformer. The lignite is pressurized by using nitrogen, which also acts as a tracer component for subsequent measurement of the synthesis gas flow. Following the lock hopper system the steam inlet is mounted. The steam supply is guaranteed by a steam generator, it is controlled with a Coriolis flow meter.

After leaving the gasifier the synthesis gas is directed into a candle filter, which removes olivine, ash and char particles. Downstream the candle filter the gas and tar analysis of the synthesis gas takes place, which is described in the following section. Afterwards a pressure-sustaining valve is mounted, which adjusts the reformer pressure. At the end of the process chain the flammable components of the synthesis gas are disposed of in the thermal afterburning.

Results

The measurement campaign in the presented work had a total duration of 72 h with 24 h of gasification operation. After reaching a safe reactor temperature above 800°C the gasification was started. During the measurement campaign five operation points were tested. Table 1 shows the main parameters of the operating points. During operation the fuel input into the gasifier as well as the steam mass flow was kept constant at each operation point while the heat duty of the combustor was adjusted to maintain steady state operation at gasification temperatures between 800 and 840°C.

Table 1: Operation points of the lignite measurement campaign

Name	Pressure [bar]	Heat duty Reformer [kW]	Heat duty furnace [kW]	Steam mass flow [kg s^{-1}]
OP1	1.3	10.8	26.2	4.0
OP2	1.3	15.1	28.9	5.7
OP3	1.5	19.4	34.0	7.3
OP4	4.0	10.8	26.8	4.0
OP4	4.0	15.1	33.2	5.7

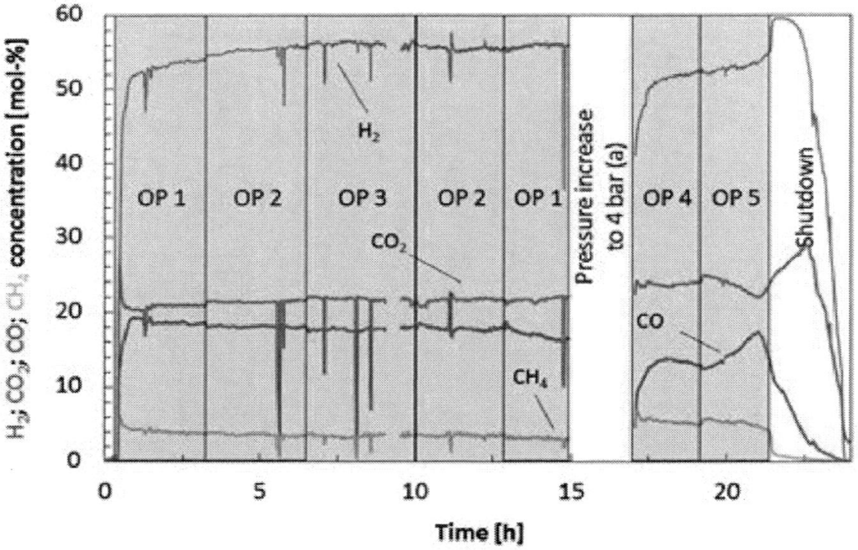

Figure 5: Chronological sequence of the gas composition during the measurement campaign

Figure 5 shows the volumetric concentration of the main gasification products, namely H_2, CO_2, CO and CH_4, as well as chronological sequence of the different operation points. The concentrations remain stable over the measurement campaign after reaching steady state after approx. 5-10 h, which can can be monitored best with the H_2 concentration. Tar content of the synthesis gas was declining with operation temperate of the gasifier and was in the range of 3-5 g/Nm³.

The gasifier showed constant hydrogen contents of 55 % in the synthesis and could safely be operated at 4 bar (a), allowing high hydrogen partial pressures for the membrane separation with nickel membranes. The medium gas composition at OP4 and OP5 was used as a bottle-mixed feed for the following membranes experiments

Membrane experiments

After single membrane experiments for preliminary tests, published in [26] and [27], the gathered information were used for a layout of a membrane bundle for the separation of approx. 1 l/min pure hydrogen. For this purpose 50 Ni 201 tubes with diameter 2.0x0.1 mm from Shaanxi Corp. were used. The following sections cover the measurement techniques for the membrane bundle experiments as well as the analysis of the influence of hydrogen separation on reaction equilibria.

Experimental setup

Figure 6 shows the membrane test rig for the membrane bundle experiments. A mass flow controller was used for the measurement of the hydrogen flow through the membrane bundle. Nitrogen acted as a tracer component to measure the dry volumetric flow rate and as a gas barrier between membrane sealing and measurement cell to prevent condensation at the temperature-sensitive sealing.

To determine the influence of the hydrogen separation on gas composition, the hydrogen permeate flow was fed back to the retentate after the measurement cell and analyzed using a µ-GC "Agilent 490" for the components CO, CO_2, N_2 and CH_4. Although measurement of H_2 is not possible as helium is used as carrier gas in the µ-GC system, it can be calculated as the difference to 100 %.

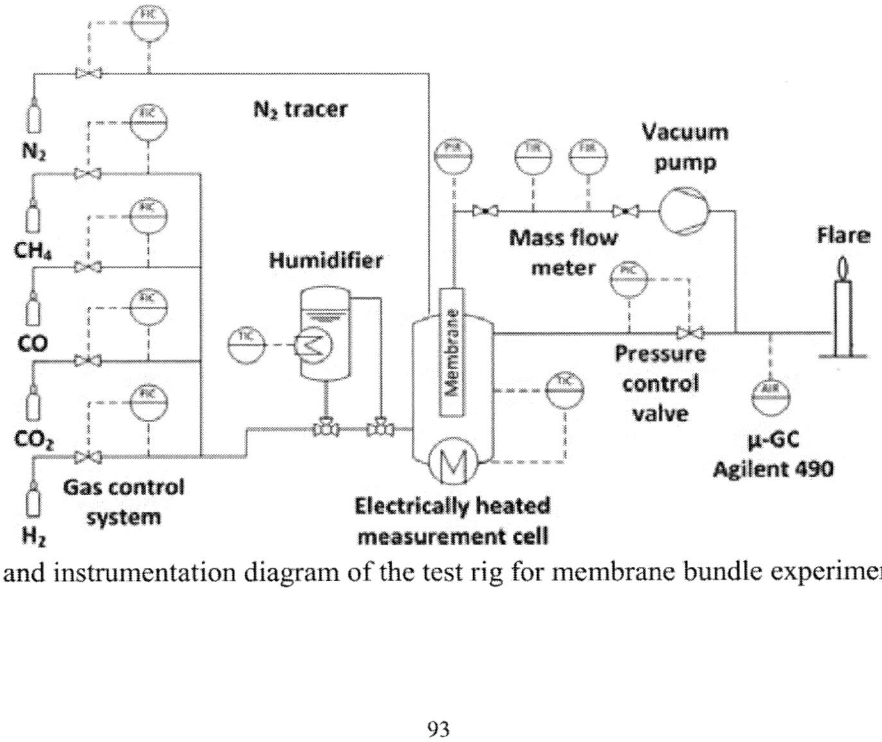

Figure 6: Piping and instrumentation diagram of the test rig for membrane bundle experiments

Results

The gas composition determined in the gasification experiments was used as feed for the measurement cell in the membrane test rig. The composition was 55 % H_2, 24 % CO_2, 16 % CO, 5 % CH_4 with a steam / ratio of 2, which resembles a steam content of 32 %. Figure 7 shows the results from these measurements. The gas composition at a hydrogen separation rate of zero does not resemble the feed composition as the gas mixture reacts due to the high temperatures and the presence of the catalytic nickel surface. Hydrogen separation has a significant influence on the turnover of CH_4 and CO. The CO turnover is negative at low hydrogen separation rates as CH_4 is converted to CO and H_2. As a result, the H_2 and CO_2 content rises due to the ongoing conversion of the feedstock. As proposed in the introduction, it is possible to influence the conversion of the synthesis gas to hydrogen by using nickel membranes.

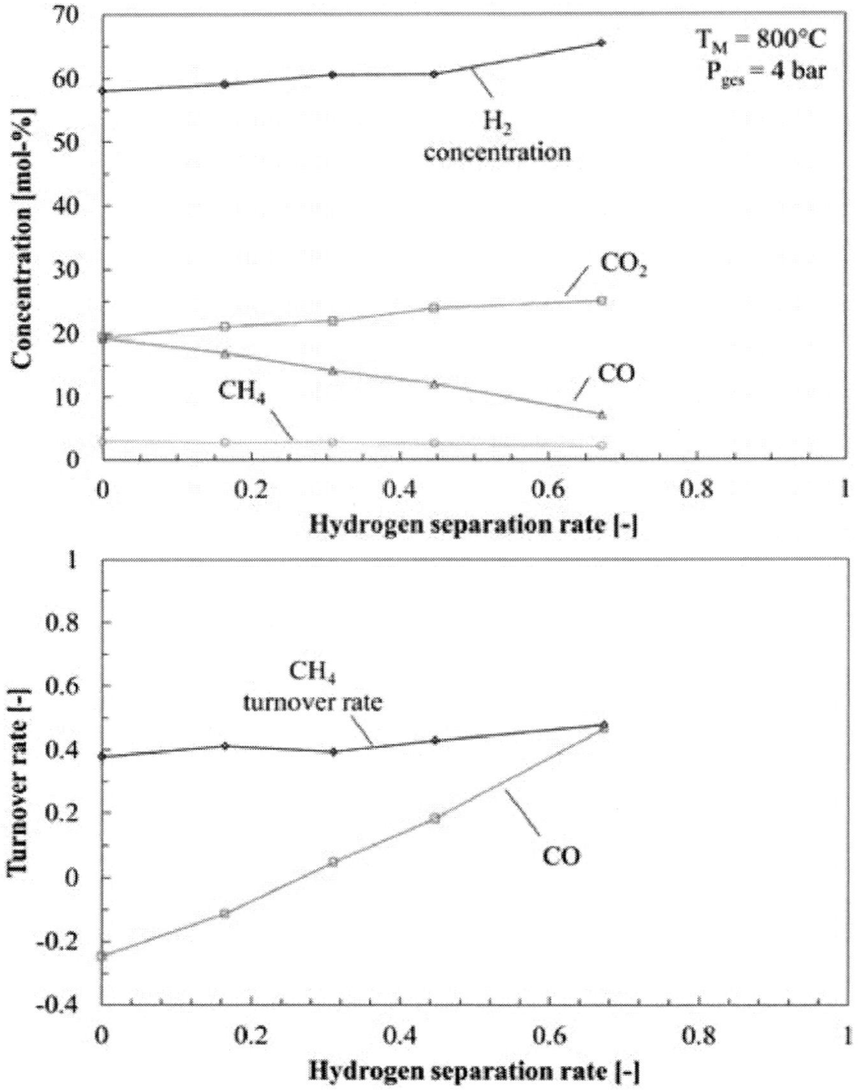

Figure 7: Resulting gas composition and turnover rates from hydrogen separation from bottle-mixed synthesis gas, Membrane temperature: 800°C, pressure 4 bar, gas composition (dry base): 55 % H_2, 24 % CO_2, 16 % CO, 5 % CH_4; steam content 32 %

Conclusion

The paper shows results from the gasification tests with the Heatpipe Reformer at FAU-EVT as well as the results from membrane experiments with a nickel membrane bundle using synthesis gas. As outlined in the state-of-the-art section, the temperature regime of gasification is very challenging for existing membrane systems such as Palladium composite membranes. The usage of nickel membranes was demonstrated and evaluated as an alternative to these systems.

Our experimental results show the good performance of the Heatpipe Reformer system and the suitability of the produced synthesis gas for hydrogen generation. The membrane results show, that the hydrogen yield and the CH_4 and CO turnover can be raised significantly by using nickel membranes.

Ongoing development at FAU-EVT concentrates on the completion and demonstration of a 100 kW SNG process chain with long term experiments. Scientific development is focused on advanced gas cleaning systems, e.g. membranes or scrubbing processes, for a sustainable usage of solid feedstock in energy process engineering

Bibliography

[1] G. Gallmetzer, P. Ackermann, A. Schweiger, T. Kienberger, T. Gröbl, H. Walter, M. Zankl, and M. Kröner, "The agnion Heatpipe-Reformer: operating experiences and evaluation of fuel conversion and syngas composition," *Biomass Convers. Biorefinery*, vol. 2, no. 3, pp. 207–215, May 2012.

[2] S. Karellas, E. Kakaras, T. Papadopoulos, C. Schäfer, and J. Karl, "Hydrogen production from allothermal biomass gasification by means of palladium membranes," *Fuel Process. Technol.*, vol. 89, no. 6, pp. 582–588, Jun. 2008.

[3] C. Schäfer, "Diffusionseigenschaften bestimmter Metalle bei der Hochtemperatur-Wasserstoffabtrennung," Technische Universität München, 2010.

[4] J. M. Leimert, M. Dillig, and J. Karl, "Hydrogen inactivation of liquid metal heat pipes," *Int. J. Heat Mass Transf.*, vol. 92, pp. 920–928, 2016.

[5] P. Quicker, "Dehydrierung von Ethylbenzol zu Styrol," FAU Erlangen-Nuremberg, 2000.

[6] H. Möllenhoff, H. J. Schröter, and H. Jüntgen, "Hydrogen Permeation through Metallic Heat Exchanger Materials during Steam Gasification of Coal," *Zeitschrift für Physikalische Chemie*, vol. 147, no. 1_2. pp. 273–284, 1986.

[7] K. Akamatsu, M. Nakane, T. Sugawara, T. Hattori, and S. Nakao, "Development of a membrane reactor for decomposing hydrogen sulfide into hydrogen using a high-performance amorphous silica membrane," *J. Memb. Sci.*, vol. 325, no. 1, pp. 16–19, Nov. 2008.

[8] Ø. Hatlevik, S. K. Gade, M. K. Keeling, P. M. Thoen, A. P. Davidson, and J. D. Way, "Palladium and palladium alloy membranes for hydrogen separation and production: History, fabrication strategies, and current performance," *Sep. Purif. Technol.*, vol. 73, no. 1, pp. 59–64, May 2010.

[9] S. Yan, H. Maeda, K. Kusakabe, and S. Morooka, "Thin Palladium Membrane Formed in Support Pores by Metal-Organic Chemical Vapor Deposition Method and Application to Hydrogen Separation," *Ind. Eng. Chem. Res.*, vol. 33, pp. 616–622, 1994.

[10] K. Kusakabe, S. Yokoyama, S. Morooka, J.-I. Hayashi, and H. Nagata, "Development of supported thin palladium membrane and application to enhancement of propane aromatixation

on Ga-silicate catalyst," *Chem. Eng. Sci.*, vol. 51, no. 11, pp. 3027–3032, 1996.

[11] E. Fernandez, J. A. Medrano, J. Melendez, M. Parco, J. L. Viviente, M. van Sint Annaland, F. Gallucci, and D. A. Pacheco Tanaka, "Preparation and characterization of metallic supported thin Pd-Ag membranes for hydrogen separation," *Chem. Eng. J.*, 2015.

[12] Y. Guo, H. Zou, H. Wu, L. Zhou, H. Liu, and X. Zhang, "Preparation of palladium membrane by bio-membrane assisted electroless plating for hydrogen separation," *Int. J. Hydrogen Energy*, vol. 39, no. 13, pp. 7069–7076, 2014.

[13] P. P. Mardilovich, Y. She, Y. H. Ma, and M.-H. Rei, "Defect-free palladium membranes on porous stainless-steel support," *AIChE J.*, vol. 44, no. 2, pp. 310–322, Feb. 1998.

[14] G. Xomeritakis and Y.-S. Lin, "CVD synthesis and gas permeation properties of thin palladium/alumina membranes," *AIChE J.*, vol. 44, no. 1, pp. 174–183, 1998.

[15] A. Basile, A. Julianelli, T. Longo, S. Liguori, and M. De Falco, "Pd-based Selective Membrane State-of-the-Art," in *Membrane reactors for hydrogen production processes*, M. De Falco, L. Marrelli, and G. Iaquaniello, Eds. Springer London, 2011, pp. 20–56.

[16] A. Kulprathipanja, G. O. Alptekin, J. L. Falconer, and J. D. Way, "Pd and Pd-Cu membranes: Inhibition of H2 permeation by H 2S," *J. Memb. Sci.*, vol. 254, pp. 49–62, 2005.

[17] H. Amandusson, L. Ekedahl, and H. Dannetun, "The effect of CO and O 2 on hydrogen permeation through a palladium membrane," *Appl. Surf. Sci.*, pp. 259–267, 2000.

[18] C. Chen, "Sulfur tolerance of Pd/Au alloy membranes for hydrogen separation from coal gas," Worcester Polytechnic Institute, 2012.

[19] U. Balachandran, T. H. Lee, C. Y. Park, J. E. Emerson, J. J. Picciolo, and S. E. Dorris, "Dense cermet membranes for hydrogen separation," *Sep. Purif. Technol.*, vol. 121, pp. 54–59, Jan. 2014.

[20] S. Y. Jeon, M. B. Choi, C. N. Park, E. D. Wachsman, and S. J. Song, "High sulfur tolerance dual-functional cermet hydrogen separation membranes," *J. Memb. Sci.*, vol. 382, no. 1–2, pp. 323–327, 2011.

[21] J. M. Polfus, W. Xing, M.-L. Fontaine, C. Denonville, P. P. Henriksen, and R. Bredesen, "Hydrogen separation Membranes based on dense ceramic composites in the La27W5O55.5–LaCrO3 system," *J. Memb. Sci.*, vol. 479, pp. 39–45, Jan. 2015.

[22] V. Gil, J. Gurauskis, C. Kjølseth, K. Wiik, and M.-A. Einarsrud, "Hydrogen permeation in asymmetric La28 − xW4 + xO54 + 3x/2 membranes," *Int. J. Hydrogen Energy*, vol. 38, no. 7, pp. 3087–3091, Mar. 2013.

[23] M. Liu, W. Sun, X. Li, S. Feng, D. Ding, D. Chen, M. Liu, and H. C. Park, "High-performance Ni–BaZr0.1Ce0.7Y0.1Yb0.1O3−δ (BZCYYb) membranes for hydrogen separation," *Int. J. Hydrogen Energy*, vol. 38, pp. 14743–14749, 2013.

[24] Z. Zhu, W. Sun, Y. Dong, Z. Wang, Z. Shi, Q. Zhang, and W. Liu, "Evaluation of hydrogen permeation properties of Ni-Ba(Zr0.7Pr0.1Y0.2)O3−δ cermet membranes," *Int. J. Hydrogen Energy*, vol. 39, no. 22, pp. 11683–11689, Jul. 2014.

[25] J. M. Leimert, P. Treiber, and J. Karl, "The Heatpipe Reformer with optimized combustor design for enhanced cold gas efficiency," *Fuel Process. Technol.*, vol. 141, pp. 68–73, 2016.

[26] J. M. Leimert and J. Karl, "Nickel membranes for in-situ hydrogen separation in high-temperature fluidized bed gasification processes," *Int. J. Hydrogen Energy*, vol. 41, no. 22, pp. 9355–9366, 2016.

[27] J. M. Leimert, A. Bartolf, and J. Karl, "Hydrogen concentration measurement in high temperature applications using hydrogen permeation," *Int. J. Hydrogen Energy*, vol. 41, no. 32, pp. 14360–14367, 2016.

469525 Performance of Transition Metal Catalysts Based on Initial Predictor for Fischer Tropsch Reaction

Abstract:

Performance of transition metal catalysts based on initial predictor for Fischer Tropsch reaction

Authors: Sumegha Godara[1], Suraj Gyawali[1] and Dr. Daniela S. Mainardi[2]

[1]Institute for Micro-manufacturing, Louisiana Tech University

[2]Institute for Micro-manufacturing, Chemical Engineering, Louisiana Tech University

All Fischer-Tropsch mechanisms known to date begin with the adsorption of carbon monoxide followed by its dissociation on a given catalysts surface. Cluster models of Co, Fe, Ni, Pd, Pt, and Ru have been explored using Density Functional Theory. Carbon monoxide adsorption energies were calculated for adsorption on all possible catalytic sites, and the most preferred CO adsorption sites were found in each case. An initial predictor that can be used to anticipate potentially effective catalysts was identified as a percentage difference, using CO adsorption energy results in combination with CO dissociation energies from the previously found sites. A greater catalysts performance is expected when that percentage difference is maximized. On pure clusters, the predictor indicates that ruthenium is expected to be the best catalyst followed by cobalt, in very good agreement with the current knowledge in this field. Hydrogen assisted CO adsorption and formation of hydrocarbons is under study.

Thus, this work presents a mechanism to quickly explore the natural potential of a catalyst material to break the CO bond. The predictor presented here can guide the synthesis of new catalysts, involving modifications of known and

currently used catalysts for the Fischer-Tropsch process, and investigations on catalysis in general.

474745 Session Keynote - Co-Extrusion: Advanced Manufacturing for Energy Devices

Abstract:

The development of mass markets for large-format batteries, including electric vehicles (EVs) and grid support, depends on both cost reductions and performance enhancements to improve their economic viability. Palo Alto Research Center (PARC) has developed a multi-material, advanced manufacturing process called co-extrusion (CoEx) to remove multiple steps in a conventional battery coating process with the potential to simultaneously increase battery energy and power density. CoEx can revolutionize battery manufacturing across most chemistries, significantly lowering end-product cost and shifting the underlying economics to make EVs and other battery applications a reality.

PARCâ€™s scale-up of CoEx for electric vehicle (EV) batteries builds on a solid base of experience in applying CoEx to solar cell manufacturing, deposition of viscous ceramic pastes, and Li-ion battery chemistries. In the solar application, CoEx has been deployed commercially at production scale where multi-channel CoEx printheads are used to print viscous silver gridline pastes at full production speeds (>40 ft/min). This operational scale-up provided invaluable experience with the nuances of speed, yield, and maintenance inherent in taking a new technology to the factory floor. PARC has leveraged this experience, adapting the CoEx process for Lithium-ion (Li-ion) battery manufacturing. To date, PARC has worked with Li-ion battery materials and structured cathodes with high-density Li-ion regions and low-density conduction regions, documenting both energy and power performance. Modeling results for a CoEx cathode show a path towards a 10-20% improvement in capacity for an EV pouch cell. Experimentally, we have realized a co-extruded battery structure with a Lithium Nickel Manganese Cobalt (NMC) cathode at print speeds equivalent to conventional roll coating processes. The heterogeneous CoEx cathode enables improved capacity in thick electrodes at higher C-rates. The proof-of-principle coin cells demonstrate the feasibility of the CoEx technology and a path towards higher energy and higher power EV pouch cells.

463732 Instability of Reversible Pump Turbines and Its Physical Origin

Abstract:

Currently, reversible pump turbines are being developed worldwide in the pumped hydro energy storage power plants. One of the paramount technical problems of the reversible pump turbines is the instabilities during the off-design working conditions. For example, in generating modes, significant pressure fluctuation has been observed in the reversible pump turbines with prominent systematic vibrations. In the present paper, several examples of the measured instabilities of reversible pump turbines are given for both the prototype and model type turbines. Based on those experimental data, the basic features of the instabilities are revealed together with the paramount influencing parameters (e.g. guide vane opening). Numerical simulations has also been performed to further investigate the physical origins of the instability. Finally, some recent progress are also briefly discussed for relieving the instability.

473631 Simplified Model of a Redox Flow Battery for Deriving Design and Dispatch Strategies in a Resource Planning Tool

Abstract:

Energy resource planning tools (ERPTs) that integrate an energy storage technology with renewable energy sources require representative models for the energy storage resources. ERPTs may involve use of optimization framework to select the optimal mix of resources and dispatch strategies. Such optimization problems can involve a yearlong time frame and several energy resources. Hence use of linear models is desirable for faster convergence of the optimization problem.

Flow battery technology has emerged as one of the potential storage candidates which can be used for stationary energy storage systems. The flexibility of scaling energy and power independently may give these systems an advantage over other energy storage technologies for applications where the energy is high and power demand is low. Flow battery systems exhibit nonlinear behavior during charging and discharging which is typical of electrochemical energy storage systems. Hence there is a need to develop a simplified (preferably linear) model without undue compromise of the underlying physics of the batteries.

Rather than modeling a flow battery as an ideal, finite source of energy with constant energy efficiency, a simple model incorporating the electrochemical behavior of a redox flow battery was developed. This model enables the optimizer to (i) choose energy and power independently (design decision) and (ii) access the performance of the chosen battery size (dispatch decision) in order to derive the optimal energy mix. The talk will also focus on quantification of errors associated with the simple model compared with the nonlinear representation.

Acknowledgements

The work presented herein was funded by the Office of Naval Research.

474162 Selection of Ion-Exchange Membrane for Vanadium Redox Flow Battery

Abstract:

The increasing share of renewable energy sources in the total energy production emphasizes the need for a reliable and cheap stationary energy storage in order to compensate for hardly predictable non-stabilities in output power of photovoltaics or wind turbines on the power transmission level. Vanadium redox flow battery (VRFB) appears as a promising solution for the stationary energy storage as it offers high efficiency (80% DC-DC), versatile arrangement of decoupled power (kW) to capacity (kWh), extended durability and fast demand response. However, the broader commercialization of the technology is still obstructed by relatively high investment costs.

Ion-exchange membrane is a key component of VRFB cell as it directly influences the power and the efficiency of the stack - the resistance of the typically used cation-exchange membrane counts for up to 50 % of the total internal resistance of the battery stack. In the same time, price of the membrane may exceed 25 % of the overall costs. Thus, the identification of highly conductive, low-cost membrane with minimal permeability for vanadium ions and sufficient durability in acidic electrolytes is vitally needed for VRFB developers.

In our contribution we present the results of a broad systematic study focused on the effect of membrane properties (charge, ion-exchange capacity, thickness, method of preparation and preconditioning) on VRFB operation. The various types of commercially available ion-exchange membranes were characterized with the respect to the properties relevant for VRFB operation such as: i) through-plane ionic conductivity in the environment of VRFB cell, ii) permeability for vanadium ions of different oxidation states, iii) performance in VRFB single-cell and iv) durability in the presence of oxidizing V^{V}electrolyte.

From the results obtained from permeation measurement it is evident that the effective diffusion coefficient of V^{3+} ions is significantly lower than of VO^{2+}ions for all the studied membranes, which corresponds with the differences in the solvation numbers of the ions. Generally higher conductivities were found for cation-exchange membranes when compared to anion-exchange ones, which is caused by the different mobility of the transported ions (hydronium vs. sulfate ions). According to our assumptions, the conductivity and selectivity of membrane significantly depend on the concentration of ion-exchange groups in the membrane. However, these properties are also influenced by the specific inner arrangement of hydrophobic and hydrophilic domains in the material. The observed trends were also confirmed by the characterization of membranes in VRFB single-cell. Based on these results we identified optimum membranes for VRFB with the respect to the energy efficiency of charging discharging cycle and maximal power output. Perfluorinated cation-exchange membranes are the solution when high power and long durability is required. Anion-exchange membranes show lower vanadium crossover and thus appears more suitable for the application when low maintenance is required.

463885 Energy Storage for the Enhancement of Penetration Level of Wind Energy: A Case Study

Abstract:

In the past five years, wind energy increases significantly in China. In terms of the equipped capacity of wind turbine, China serves as the top country in the world. However, serious problems occur due to the connection of the wind energy with the electrical grid. For example, the fluctuation nature of the wind energy causes significant instabilities in the grid, leading to the rejection of the wind energy. In some local area of China, the rejection rate of the wind energy could reach 40% or even more. To solve this dilemma, it is quite necessary to develop the energy storage facilities (in terms of both large-scale and small-scale capacities). In the present paper, effects of the energy storage on the penetration level of wind energy are investigated based on the wind farms in the western parts of the Inner Mongolia province of China. The detailed analysis are given based on the quantitative data obtained from the government and industries. The importance of the energy storage are categorized into several groups and several emerging trends are briefly discussed.

474663 An Investigation of a Polysulfide - Polyiodide Aqueous Redox Flow Battery

Abstract:

Redox flow batteries (RFBs) have attracted considerable academic and industrial interest based on their favorable combination of performance and safety in the widespread integration of renewable, non-dispatchable energy sources and in the improvement of energy efficiency of the electric grid. However, at present, RFBs are broadly considered too expensive for widespread deployment. A cost-effective aqueous RFB based on alkaline metal polysulfide (PS) and polyiodide (PI) is shown here for the application grid-scale energy storage. The proposed RFB applies earth-abundant elements, which shows great potential to meet the stringent cost target for RFBs ($100/kWh). The OCV of the proposed RFB is ~ 0.88 V with a theoretical energy density as high as ~ 46 Wh/L that is ~84% higher than that of all vanadium redox flow batteries. Galvanostatic and potentiostatic studies were carried out on both the negative and positive sides, which sheds light on the electrochemistry of sulfide and iodide species, respectively. The cell performance of the proposed PS-PI chemistry was further studied using a flow battery architecture operated at 5 mL/min and 20 degree Celsius. The resultant RFB displayed stable performance at 15 mA/cm^2 for over 150 hours (25 cycles) with relatively small capacity decay. A system diagnosis after the cycling experiment revealed that membrane (sodiated Nafion membrane) degradation is the primary reason for the significantly decreased energy efficiency of the cell after 25 cycles.

474797 Progress in Electro Acoustic Signal Analysis

Abstract:

We have recently determined a correlation between the acoustic time of flight and intensity and the structure, state of charge, and state of health of closed system electrochemical energy storage. We have shown that regardless of chemistry and geometry, an acoustic time of flight signal can be related to these properties of the cell.

Because a closed cell is effectively a mass redistribution reactor, and in a standard cell the volume is effectively fixed, the distribution of density within a battery must change as a function of state of charge and, along with density, the bulk moduli of the anode and cathode changes as well. The shifts in density and modulus also change the acoustic attenuation in a battery

In this presentation we will review the physical basis of our hypothesis and present progress in 1) correlating structural evolution and failure analysis with acoustic signal evolution, 2) present preliminary inverse models which describe the changes we see in the acoustic/state of charge response and 3) present â€œphysics freeâ€ machine learning results to examine the uniqueness of an acoustic signal for a given battery in a given state of charge and health.

463541 Model-Assisted Development of Microfabricated 3D $Ni(OH)_2$ Electrodes with Rapid Charging Capabilities

Abstract:

Model-assisted Development of Microfabricated 3D $Ni(OH)_2$ Electrodes with Rapid Charging Capabilities

Recently, several studies have focused on the development of materials that realize batteries with improved power densities, and thus, rapid charging and discharging capabilities. Yet, they are primarily based on non-deterministic fabrication techniques which limit the control over critical electrode dimensions affecting the power performance of the battery, such as the electrode surface area and the active material thickness. The study reported herein involves rationally designed and deterministically engineered 3D structures with nearly precise control over the optimized dimensions that address the aforementioned critical factors affecting the high-power capability and ultimately determine the performance of the electrochemical system.

A series of fabrication methods encompassing electrochemical techniques and microfabrication technologies have been utilized for the formation and characterization of the scalable, well-ordered, and high-surface-area 3D architectures that can be potentially used as high-power electrodes in a variety of applications, ranging from autonomous microsystems to macroscale portable electronics. These versatile structures were based on a large number of thin layers of two different materials, i.e., Cu and Ni, the former being sacrificial while the latter was structural, fabricated through a temporary photoresist mold in an alternating sequence via an automated robotic system. The final structure was obtained by the selective removal of the sacrificial Cu layers, leaving behind a large number of separate Ni layers to be coated with the electrochemically active material. Nickel hydroxide ($Ni(OH)_2$), one of the most well-studied active material in secondary (i.e., rechargeable) battery systems along with Li-ion chemistry, was selected as the active material to be deposited onto the 3D multilayer structure. Electrodes with areal capacities as high as 2.43 mAh cm^{-2}

were realized, which were able to deliver more than 50% of their capacities when charged at ultrafast rate of 150C. The resultant electrodes also featured remarkable cycling stability when charged and discharged at high rates for more than 80 cycles.

For the optimization of the characteristic dimensions of the electrodes, a two-dimensional mathematical model has been developed in COMSOL 5.2 by employing fundamental mass transport and reaction kinetics principles. The geometry of a repeating unit of the multilayer electrode structure was studied. Fickâ€™s second law, Nernst-Planck equation, and Butler-Volmerâ€™s equation were used to describe the solid phase diffusion, the electrolyte phase transport, and the electrode/electrolyte interfacial kinetics, respectively. The model was shown to successfully predict the energy/power tradeoff of the fabricated multilayer electrodes. Diffusion coefficient of protons within the $Ni(OH)_2$ active material, a critical parameter that determines the power performance of the $Ni(OH)_2$ electrode, was experimentally measured using cyclic voltammetry. The obtained value of $2\sim3*10^{-10}$ cm^2/s is within the range reported from existing literatures and close to the best-fit value ($3.4*10^{-10}$ cm^2/s) obtained from the model. Accordingly, effort was directed towards finding the optimal electrode dimensions that lead to enhanced power performance based on the model. Geometrical parameters that were taken into account included the active material thickness, the interlayer spacing between two individual layers, and the nickel current collector thickness. Projections showed that multilayer structures based on the optimized dimensions exhibited superior performance than the experimentally fabricated non-optimized structures. Depending on the energy and power needs, this model opened up the possibility for the realization of deterministically engineered electrodes for both micro- and macroscale applications.

464527 Electrode-Supported Macroporous Ceramic Membrane Separator for Lithium Ion Batteries

Abstract:

The emergence of lithium-ion batteries as the most popular energy storage system is mainly due to their high energy density and long cycle life. They find application in a wide variety of portable electronic devices as well as electric vehicles and smart grids. However, the relatively frequent occurrences of battery fires and explosions are a major cause of concern. A significant source of this flammability, along with the electrolyte, is the polymeric separator used widely in commercial lithium-ion batteries. Lithium ion batteries with inorganic separators offer advantages of safer operation. An inorganic macroporous membrane separator for lithium-ion battery was prepared by blade coating an α-Al_2O_3 slurry directly on the electrode followed by drying. The improved separator preparation involves a twice-coating process instead of coating the slurry all at once in order to obtain a thin (~40μm) and uniform coat. Unlike state-of-the-art polyolefin separators such as polypropylene (PP) which are selectively wettable with only certain electrolytes, the excellent wettability of α-Al_2O_3 allows the coated separator to function with different electrolytes. The coated α-Al_2O_3 separator has much higher resistance to temperature effects than polyolefin counterparts. The inorganic coated separator retains its dimensional integrity at temperatures as high as 200ºC, thus eliminating the possibility of a short circuit during thermal runaway. Lithium ion batteries half-cells and whole cells with coated α-Al_2O_3 separator exhibit electrochemical performance comparable with that with PP separator at room temperature. However, the coated separator shows better cycling performance than PP separator under extreme temperatures when tested over a wide temperature range of -30°C to +60°C. Therefore, the coated α-Al_2O_3 separator is very promising for uses in safe lithium-ion batteries for energy storage applications.

465548 First-Principles Density Functional Theory Modeling of Li Binding: Thermodynamics and Redox Properties of Quinone Derivatives for Lithium-Ion Batteries

Abstract:

First-Principles Density Functional Theory Modeling of Li Binding: Thermodynamics and Redox Properties of Quinone Derivatives for Lithium-Ion Batteries

Ki Chul Kim,[1] Tianyuan Liu,[2] Seung Woo Lee,[2] and Seung Soon Jang[1]

[1]*School of Material Science and Engineering, Georgia Institute of Technology, 771 Ferst Drive NW, Atlanta, GA 30332-0245, USA*

[2]*G. W. Woodruff School of Mechanical Engineering, Georgia Institute of Technology, 801 Ferst Drive NW, Atlanta, GA 30332-0405, USA*

Abstract

The Li-binding thermodynamics and redox potentials of seven different quinone derivatives are investigated to determine their suitability as positive electrode materials for lithium-ion batteries. First, using density functional theory (DFT) calculations on the interactions between the quinone derivatives and Li atoms, we find that the Li atoms primarily bind with the carbonyl groups in the test molecules. Next, we observed that the redox properties of the quinone derivatives can be tuned in the desired direction by systematically modifying their chemical structures using electron-withdrawing functional groups. Further, DFT-based investigations of the redox potentials of the Li-bound

quinone derivatives provide insights regarding the changes induced in their redox properties during the discharging process. The redox potential decreases as the number of bound Li atoms is increased. However, we found that the functionalization of the quinone derivatives with carboxylic acids can improve their redox potential as well as their charge capacity. Through this study, we also determined that the cathodic activity of quinone derivatives during the discharging process relies strongly on the solvation effect as well as on the number of carbonyl groups available for further Li binding.

466882 Conductivity Degradation of Polyvinylidene Fluoride Binder during Cycling: Measurements and Simulations for Lithium-Ion Batteries

Abstract:

Battery electrodes are complex multiphase composites which must provide efficient bicontinuous networks for transport of electrons (through the particle phase) and positive lithium ions (through the electrolyte filled pores of the electrode). A crucial but often neglected element of battery electrodes is the binder, typically a mixture of polyvinylidene fluoride (PVDF) and carbon black. The binder has two primary roles â€" to provide mechanical integrity and to improve electrical conduction of the electrodes. Migration of the binder has also been implicated as a potential mechanism of capacity fade in rechargeable lithium ion batteries.

We will present experimental characterization of the polymer binder for battery applications. Mechanical properties of the composite binder will be shown for both dry films and also binder swollen with carbonate electrolytes used in rechargeable lithium batteries. The electrical properties are strongly dependent on the applied stress. The evolution of mechanical and electrical properties of the binder after repeated cycling will be shown.

Then we will examine the impact of cyclic mechanical stresses on the performance of lithium cobalt oxide cathodes. Experimental results will be presented on the evolution of electrical conductivity as a function of mechanical cycling. Mechanisms for electronic conductivity in these complex porous composite electrodes will be investigated using mesoscale simulations using experimentally determined three dimensional structures of battery cathodes. Implications for battery internal resistance and cycling stability will be discussed.

Sandia National Laboratories is a multi-program laboratory managed and operated by Sandia Corporation, a wholly owned subsidiary of Lockheed Martin Corporation, for the U.S. Department of Energyâ€™s National Nuclear Security Administration under contract DE-AC04-94AL85000. SAND2016-4371 A

474683 Polymeric Electrolyte Additives for Suppressing Zinc Dendrite Formation in Rechargeable Batteries with Zinc Anodes

Abstract:

Rechargeable zinc-air batteries are attractive because they utilize non-flammable aqueous electrolytes and offer high energy densities. However, a major roadblock in the development of rechargeable zinc-anodes is the dendritic morphology evolution during zinc electrodeposition (i.e., during battery charging). Zinc dendrites are a major source of capacity loss and early cell failures.

In the present talk, we will demonstrate the use of polymeric additives, such as polyethylenimine, as electrolyte additives for effectively suppressing dendrite formation during zinc electrodeposition. Dendrite suppression by additives will be characterized using in situ optical microscopy. Steady-state and transient electrochemical polarization measurements on a rotating disk electrode, combined with electrochemical quartz crystal microgravimetry will be used to unravel the mechanism by which additives suppresses zinc dendrite growth. Additive injection studies implemented in a rotating disc electrode configuration, coupled with diffusion-adsorption modeling, will be discussed as a reliable technique for precisely measuring the diffusion coefficient and adsorption rate constant of aforementioned polymeric additives. These additives' properties provide insights into the rate-controlling processes responsible for additives-assisted zinc dendrite suppression.

475724 High-Efficiency Electrochemical Conversion of CO_2 to Value Added Fuels Current Status, Challenges, and Future Directions

Abstract:

The electrochemical reduction of carbon dioxide (CO_2) into liquid fuels especially coupling with the intermittent renewable electricity offers a promising means of storing electricity in chemical form, which reduces the dependence on fossil fuels and mitigates the negative impact of anthropogenic CO_2 emissions on the planet. Although converting CO_2 to fuels is not in itself a new concept, the field has not substantially advanced in the last 30 years primarily because of the challenge of discovery of structural electrocatalysts and the development of membrane architectures for efficient collection of reactants and separation of products. In this presentation, I would summarize recent advances in electrocatalytic conversion of CO_2 and presents the challenges and future directions in producing value-added fuels.

Currently, copper, silver and tin are the most extensively scrutinized electrocatalysts, which can convert CO_2 to CO, CH_4, C_2H_4, alcohols and carbolic acid in aqueous electrolytes. A long-lasting challenge in the electrosynthesis of fuels from CO_2 is to achieve a durable electrochemical performance, an equally important characteristic of an electrode as the activity and selectivity, but it has not yet been clearly addressed in the open literature. Hypotheses for the origin of the deactivation of CO_2 reduction electrode included the formation of intermediates like oxide, hydroxide and carbon species, as well as the accumulation of a soluble CO_2 reduction product in the electrolyte causing a remarkable decrease in the selectivity. Hori et al. examined the deactivation of Cu electrode and postulated that the degradation of the cell performance was originated from the deposition of iron and zinc contaminants in the electrolyte on the active surface of the catalyst. The impurity alone, however, cannot explain the degradation since the ultrapure electrolyte coupled with pre-electrolysis was used, in which the degradation was still observed. I will discuss

approaches to simultaneously achieve high efficiency and high stability of an electrocatalyst for the conversion to CO_2 to value-added fuels.

456332 Design and Economic Analysis of IGCC Power Plants with CO_2 Capture

Abstract:

Power producing industry has largely relied on coal as a fuel source and will continue to depend on it for coming decades under the current situation. However, in order to curtail CO_2 emissions while keeping high efficiency and economic feasibility of the processes is a key challenge. Post combustion coal-fired power plants usually have lower capital costs and cost of electricity without CO_2 capture. On the other hand, pre-combustion plants tend to be less expensive if the current CO_2capture systems are to be added. This analysis suggests that pre-combustion technology for the power generation can be an attractive option if carbon capture and storage (CCS) technology is implemented on a large scale. In this study IGCC process has been simulated in order to analyze the overall plant output with CO_2 capture using a physical solvent. Three design schemes have been developed to analyze the overall plant performance and economics. In order to ensure a fair evaluation of analysis, a consistent and transparent methodology has been followed for all the cases. First two cases use the water gas shift reactions scheme with sour shift catalysis process. The resulted syngas free of CO_2 can either be combusted by using air or O_2. The first case uses air as an oxidant for burning H_2 and the combustor temperature is controlled by air as well. In the second case, O_2 is used as an oxidizing agent in the H_2 combustion. However, combustion temperature control is achieved by recycling the captured CO_2. In the third model, WGS reactor has been removed. The syngas composed of CO and H_2 is sent directly for combustion which makes it similar to the oxy-fuel combustion process. In this way, captured CO_2 can be recycled back to control the combustor temperature in 2nd and 3rdcase. The results showed that the net power generation capacity and efficiencies calculated for case 1, case 2 and case 3 are 375Â MWe, 354 MWe, 388 MWe and 35.87%, 33.89%, 37.15% respectively. Finally, an economic analysis has been performed which shows thatÂ the total investment cost

(TIC) and operation and maintenance (O&M) required per unit (MW) of electricity generated for case 1, case 2 and case 3 is (2.97, 3.40, 2.64) and (0.2486, 0.2706, 0.2242) Mâ,¬â€™s respectively. With the highest net power generation capacity and least TIC and O&M cost requirements, case 3 has been evaluated as the most feasible design. If the sole purpose of IGCC power plant is power generation with CO_2 capture, case 3 design offers a cost reduction of 8.17% and 6.59% in the total investment and O&M cost requirements respectively as compared to the base case design.

468074 Optimization of the CCS Network Under Harsh Geographical Conditions, Using GIS and MINLP Modeling

Abstract:

With the evermore increasing amount of greenhouse gas emissions every year, various efforts to mitigate its amount have earned worldwide attention. Carbon capture and storage, a practical technology to achieve this task, has been studied from different viewpoints, namely capture, transportation, and utilization of CO2. While the CO2 capture technology has been studied very deeply, studies for practical implementation of the CO2 transportation infrastructure has received relatively less attention. Although the CO2 capture technology is the most critical from an economic point of view, research on the topic of economic transportation of CO2 is essential for commercialization of the CCS technology, since mass amounts of it have to be transferred across long distances. For mass amounts of CO2 transport, pipelines are deemed as the most economical option. Previous studies concerning this problem have introduced the method of pipeline network optimization regarding the geographical conditions to minimize the cost of nation-wide transport. These studies used a combination of geographical information systems and mixed-integer program (MIP) modeling to optimize the network. However, geographical conditions vary greatly according to the region of pipeline installation, and linear pipeline models are inadequate in rigorously evaluating the cost of pipelines. In this study, a pipeline network for transporting mass amounts of CO2 in harsh geographical conditions was optimized, using mixed-integer nonlinear programming (MINLP). A commercial GIS software ArcGIS was used to obtain the least cost paths from the CO2 sources to the reservoirs, then the cost factor for penalizing the pipelines built on a certain plain was segmented into more rigorous levels. This data was transferred into GAMS, an optimization software and designed into a MINLP problem, using a weight-based cost model for evaluating the pipeline costs, for more accurate calculations. A case study in South Korea was conducted to test the effect of rigorous geographical penalization and cost model usage. The results of the study showed that a more rigorous network can be obtained by implementing more segmented penalty factors corresponding to the varying elevation. Also, it showed that rather than avoiding the rough landscape entirely, it may be more economical to gather the emitted CO2 from nearby sources into local junction nodes, then transport it through mainstream pipelines to the reservoir.

AMINE-BLENDS SCREENING AND CHARACTERIZATION FOR CO_2 POST-COMBUSTION CAPTURE

*Abdullah Al Hinai, Dang Viet Quang, and Mohammad R.M. Abu Zahra**
Department of Chemical and Environmental Engineering, Masdar Institute of Science and Technology, P.O. Box 54224, Masdar City, United Arab Emirates
** Tel.: +97128109181 - email: mabuzahra@masdar.ac.ae*

Abstract

This work studied the performance of novel amine blends in order to find a blend of alkanolamines, which possess all the required properties for efficient CO_2 capture. 2-(Butylamino)ethanol (2BAE), a secondary amine was selected to blend with three tertiary amines including 2-(dimethylamino)ethanol (2DMAE), N,N,N′,N′-Tetramethyl-1,3-propanediamine, (TMPDA), and N-methyldiethanolamine (MDEA). Different blends with a total amine concentration of 30 wt% were tested at 40°C using a flow micro reaction calorimeter to determine CO_2 loading and heat of absorption. Obtained results reveals that blending 2BAE with tertiary amines results in the increase of absorption heat compared to single tertiary amines. 2BAE/TMDAP blend has the highest CO_2 loading; 5, 10, and 15 wt% of 2BAE blending with 25, 20 and 15 wt% of TMDAP absorbed 1.09, 0.82 and 0.74 mole CO_2/mole amine, respectively. Results from CO_2 loading, absorption heat, and absorption kinetics revealed that blend containing 5wt% of 2BAE with TMDAP 25wt% or 2DMAE 25wt% provides the best CO_2 absorption performance.

Keywords: CO_2 capture; Absorption; CO_2 loading; Post-combustion; Amine blend;

Introduction

Concerns about global warming and climate change have urged scientists to find more effective methods to reduce anthropogenic CO_2 emission. Global CO_2 emission has increased significantly in recent years [1]. This causes great increase in atmospheric CO_2 concentration, which has reached 402 ppm in August 2016 [2]. Fossil fuel accounts for about 90% of total global CO_2 emission in which fossil fuel fired power plants are considered to be major sources of CO_2 emission [1]. Many efforts have been taken to reduce CO_2 emission in which carbon capture and storage (CCS) has been considered to be the most viable technology for CO_2 reduction. Three different CO_2 capture concepts have been introduced including pre-combustion capture, oxy-fuel combustion capture and post-combustion in which post-combustion capture is the feasible because it can be easily retrofit to the existing fleet of fossil fuel fired power plants [3]. The post combustion capture using amine solvents is the most major technology for CO_2 capture, which involves a chemical absorption of CO_2 into aqueous amine like ethanolamine [4-7]. Many works have been carried out to demonstrate the feasibility of this technology at pilot scale [8-11]. The application of this technology, however, is hindered by high energy consumption for the regeneration process, degradation, and corrosion. However, capturing CO_2 using amine solvent is still attractive due to its high absorption kinetics, which allow this technology to capture CO_2 from various sources including diluted CO_2 gas stream.

Various amine compounds including primary, secondary, and tertiary amines have been investigated for CO_2 capture application in which monoethanolamine (MEA), diethanolamine (DEA), and methyldiethanolamine (MDEA) are the most widely used amines for CO_2 absorption . The primary and secondary alkanolamines have a high reaction rate upon CO_2 absorption, which begins with the formation of a zwitterion (1) that then transfers a proton to an amine to form carbamate (2). Their maximum possible CO_2 loading is 0.5 mole of CO_2/mole of amine (two amine molecules react with one CO_2 molecule). At high pressure, the carbamate could be hydrolysed to form a free amine and bicarbonates, and then the free amine will again react with CO_2 (3).

$$R_1R_2NH + CO_2(aq) \leftrightarrow R_1R_2NH^+COO^- \quad (1)$$
$$R_1R_2NH + R_1R_2NH^+COO^- \leftrightarrow R_1R_2NH^+ + R_1R_2NCOO^- \quad (2)$$
$$R_1R_2NCOO^- + H_2O \leftrightarrow R_1R_2NH + HCO_3^- \quad (3)$$

For tertiary amines, the maximum CO_2 loading could be 1 mole of CO_2 per mole of amine (one molecule of amine reacts with one molecule of CO_2) with the formation of bicarbonates (4). Although the tertiary alkanolamines have higher CO_2 loading, they have low reactivity with CO_2 in comparison with the primary or secondary amines.

$$R_1R_2R_3 + CO_2(aq) + H_2O \leftrightarrow R_1R_2R_3NH^+ + HCO_3^- \qquad (4)$$

Conventionally, aqueous MEA solution 30 wt% has been widely investigated and used for CO_2 capture [12]. However, a MEA based CO_2 capture process is very energy intensive due to the high energy requirement for MEA regeneration and this prevents this conventional post-combustion process from being used for industrial applications [13]. Recently, great effort has been spent to find alternative amines that may help increase CO_2 absorption capacity and reaction kinetics while lower regeneration energy. Amine system associated with high CO_2 absorption capacity, low regeneration energy, and high reaction kinetics results in lower cost of CO_2 capture. It is not easy to find a single amine that satisfies all above requirements. Our previous study did a survey on a large number of primary, secondary and tertiary amines bearing different functional groups and found that the primary and secondary amines have fast reaction kinetics but high absorption heat, whereas tertiary amines possess low absorption heat but slow reaction kinetics [14]. To take the advantage of both amine types, an idea to combine these amines in an amine blend system has been considered.

Following a previous experimental work which shows 2-(Butylamino)ethanol (2BAE) as a suitable amine for carbon capture as it is a secondary amine with a CO_2 loading of 0.73 mole CO_2/mole amine and a heat of absorption of -74.38 kJ/mol of CO_2. Tertiary amines including 2-(dimethylamino)ethanol (2DMAE), N,N,N′,N′-Tetramethyl-1,3-propanediamine, (TMPDA), and N-methyldiethanolamine (MDEA) have high CO_2 loading but slow reaction kinetics have been selected to blend with 2BAE. For a total of 30 wt% of aqueous amines, the 2BAE concentration is varied from 0 to 15 wt% while the respective tertiary amine contributes the remaining part to make a total of 30 wt%. CO_2 absorption performance and heat of absorption were investigated using a flow Micro Reaction Calorimeter (URC) to evaluate the effectiveness of the blends.

Experimental

Material

2-(Butylamino)ethanol, 98% (2BAE), 2-(dimethylamino)ethanol, 98% (2DMAE), N,N,N′,N′-Tetramethyl-1,3-propanediamine, 99% (TMPDA), N-methyldiethanolamine, 99% (MDEA) were purchased from Sigma Aldrich and used without further purification. CO_2 gas (99.9 %) was supplied by Gulf Industrial Gases CO. L.L.C.

Determination of CO2 loading and absorption heat

A description of the equipment can be found in the previous study [14]. A flow micro-reaction calorimeter that was supplied by Thermal Hazard Technology (UK) was used to evaluate CO_2 loading and the absorption heat for aqueous amine solutions (enthalpy of reaction/ΔH (kJ/mole of CO_2)) (Figure 1 (a)). The calorimeter is controlled by URC control software from Thermal Hazard Technology. In a typical experiment, a sample-containing cell was placed inside the calorimeter, and the temperature was set to 40 °C and atmospheric pressure. As the experiment starts, the variation in power (mW) over time (s) is recorded by the software. When the signal is stable, CO_2 gas is introduced into the cell at a rate of 0.5 ml/min (± 0.01 ml/min). The CO_2 flow is controlled by a mass flow controller, and the flow passes through a desiccant column to remove moisture before entering the sample cell. Because the CO_2 reaction with the aqueous amine solution is exothermic, the power signal initially increases, and then the signal decrease becomes constant when the reaction is complete (equilibrium) (Figure 11 (b)). The CO_2 loading (mole CO_2/mole amine) was calculated by dividing the mole of CO_2 absorbed per the mole of lean amine as described in equation 5.

$$CO_2\text{ loading } = \frac{(m_1 - m_{0)/44}}{m_0 * (\frac{C1}{M1} + \frac{C2}{M2})} \qquad (5)$$

Where, m_0 is the mass of lean amine and m_1 is the mass of CO_2 loaded amine that is the mass difference between the analysis cell containing amine after CO_2 adsorption and empty cell. C1 and M1 are concentration and molecular weight of amine 1 in the blend and C1 and M2 are concentration of amine 2 in the blend.

The integral heat Q (in kJ) was determined by using URC analysis software. The heat of absorption (ΔH - kJ/mole of CO_2) was calculated by dividing integral heat Q by the moles of absorbed CO_2. The absorption heat is proportional to amount of CO_2 absorption, therefore accumulative absorption heat evolved is used to monitor the CO_2 absorption progress and CO_2 absorption kinetics.

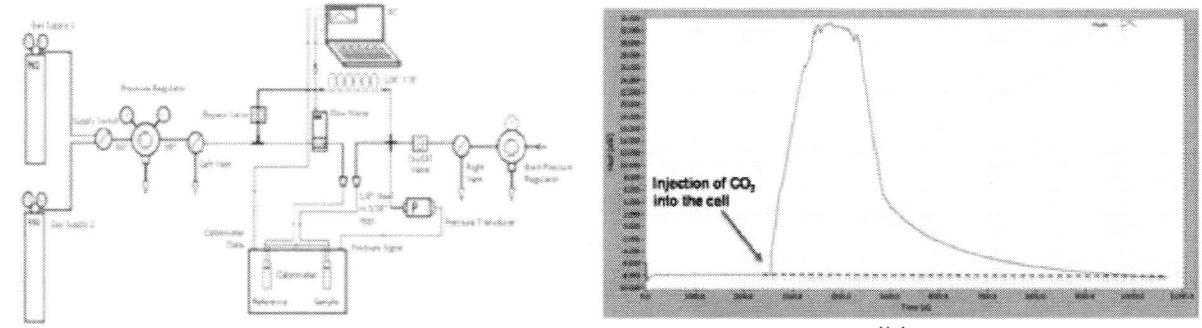

(a) (b)

Figure 1. Schematic of the flow Micro Reaction Calorimeter (a) and the data obtained from this equipment (b) [14].

Results and discussion

Previous study screened numerous amines and indicated that 2BAE is a promising candidate for CO_2 capture [14]. Aqueous 2BAE 30 wt% solution has CO_2 loading of 0.73 mole CO_2/mole amine and absorption heat of 74.4 kJ/mole CO_2. This is better than conventional MEA solution at the same concentration, which has CO_2 loading of 0.69 mole CO_2/mole amine and absorption heat of 85 kJ/mole CO_2. Three tertiary amines with different properties have been selected for blending with 2BAE including TMDAP, 2DMAE, and MDEA. These tertiary amines have lower absorption heat compared to 2BAE, however, their CO_2 loading ranges from lower to higher than that of 2BAE. TMDAP, 2DMAE, and MDEA have CO_2 loading of 1.32, 0.77, and 0.52 mole CO_2/mole, respectively. CO_2 loading of TMDAP is much higher, while CO_2 loading of MDEA is lower than that of 2BAE, meanwhile CO_2 loading of 2DMAE is similar to that of 2BAE. Experimental CO_2 loading and absorption heat of 2BAE blended with 3 tertiary amines are presented in Table 1.

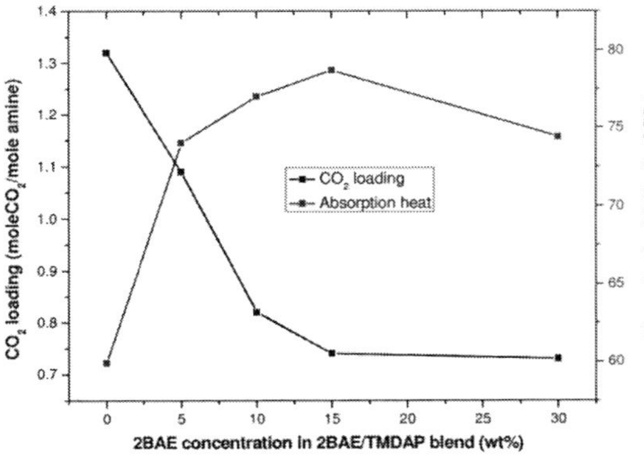

Figure 2. Variation of CO_2 loading and absorption heat as a function of 2BAE concentration in 2BAE/TMDAP blend

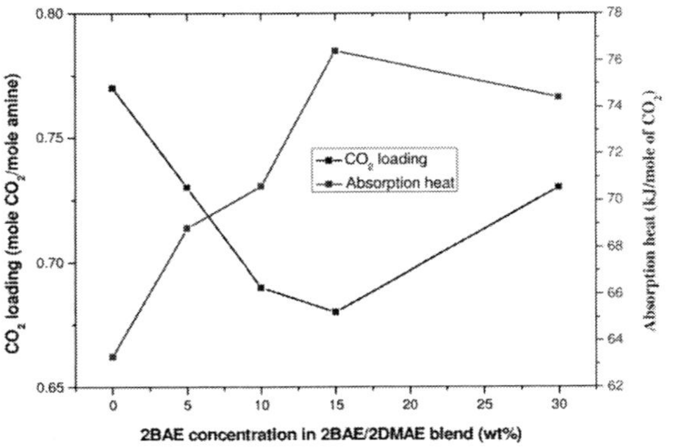

Figure 3. Variation of CO_2 loading and absorption heat as a function of 2BAE concentration in 2BAE/2DMAE blend

For 2BAE/TMDAP blend, the absorption heat of blend increased, whereas the CO_2 loading decreased compared with that of single TMDAP amine. CO_2 loading of TMDAP reduced from 1.32 to 1.09 mole CO_2/mole amine as 5 wt% of TMDAP was replaced by 2BAE and the CO_2 loading of blend is equal to that of single 2BAE as the 15 wt% of TMDAP was replaced with 2BAE (50:50 weight ratio). The result reveals that blend is efficient only when the 2BAE/TMDAP ratio is below 50%. Figure 2 shows variation of CO_2 loading and absorption heat as a function of 2BAE concentration in the blend and indicates that the best combination between 2BAE and TMDAP is 5 wt% of 2BAE and 25 wt% TMDAP in total 30 wt% amine in solution.

Table 1. CO_2 loading and absorption heat of amine blends. Tests were conducted at 40 °C using pure CO_2 at atmospheric pressure

	2BAE/TMDAP Blends				
2BAE concentration (wt%)	30	15	10	5	0
TMDAP concentration (wt%)	0	15	20	25	30
Heat of absorption (kJ/mole CO_2)	74.4	78.7	77.0	74.0	59.9
CO_2 loading (mole CO_2/mole amine)	0.73	0.74	0.82	1.09	1.32
	2BAE/2DMAE Blends				
2BAE concentration (wt%)	30	15	10	5	0
2DMAE concentration (wt%)	0	15	20	25	30
Heat of absorption (kJ/mole CO_2)	74.4	76.4	70.6	68.8	63.3
CO_2 loading (mole CO_2/mole amine)	0.73	0.68	0.69	0.73	0.77
	2BAE/MDEA Blends				
2BAE concentration (wt%)	30	15	10	5	0
MDEA concentration (wt%)	0	15	20	25	30
Heat of absorption (kJ/mole CO_2)	74.4	70.9	65.1	58.8	52.8
CO_2 loading (mole CO_2/mole amine)	0.73	0.59	0.57	0.55	0.52

Blending 2BAE with 2DMAE also results in increasing absorption heat and decreasing CO_2 loading of 2DMAE, but the change is not significant; 5 wt% of 2BAE in the blend caused reduction in CO_2 loading of 2DMAE from 0.77 to 0.73 mole CO_2/mole amine and raising absorption heat to 68.8 from 63.3 kJ/mole CO_2. Figure 3 reveals that replacing from 5 to 10 wt% of 2DMAE with 2BAE in 2DMAE 30 wt% solution will result in the best blend in term of CO_2 loading and heat of absorption. Contrast to TMDAP and 2DMAE, the addition of 2BAE in to MDEA result in increasing CO_2 loading of the blends at all 2BAE/MDEA ratios. The higher CO_2 loading of blend in comparison with single MDEA is contributed by 2BAE, whose CO_2 loading is higher than MDEA.

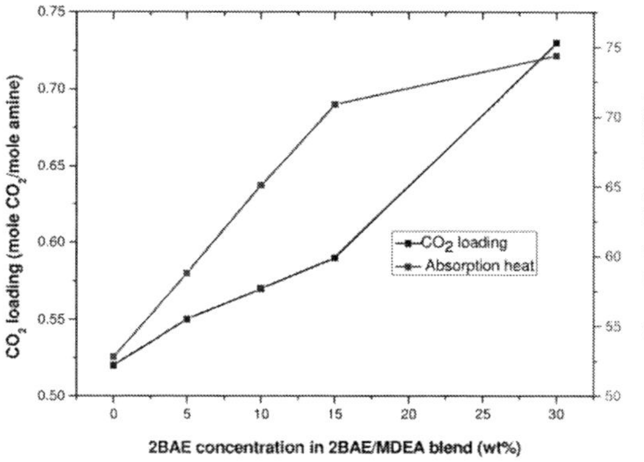

Figure 4. Variation of CO_2 loading and absorption heat as a function of 2BAE concentration in 2BAE/MDEA blend

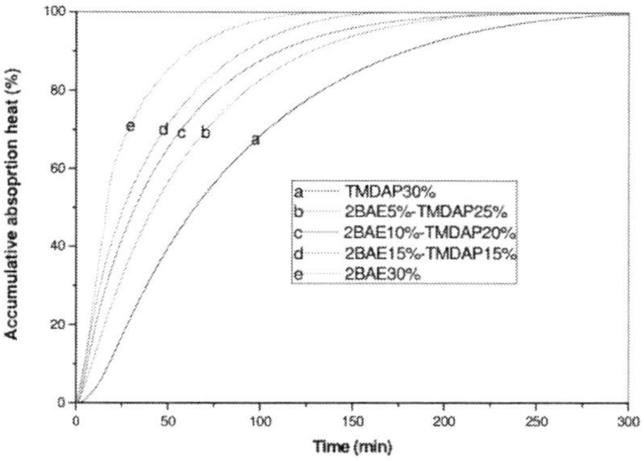

Figure 5. CO_2 absorption kinetics of 2BAE/TMDAP blends

Slow CO_2 absorption kinetics of tertiary amines is expected to be improved by blending with primary or secondary amine. In fact, a blend with low absorption heat and high CO_2 loading but slow absorption rate is still not realistic to CO_2 capture. Therefore, in addition to CO_2 loading and absorption heat, the CO_2 absorption kinetics should be investigated. In this study, heat evolved during CO_2 absorption recorded by flow micro calorimeter is proportional to amount of CO_2 absorbed. To interpret this data to kinetics, the percentage of accumulative absorption heat over absorption time was used. The CO_2 absorption kinetics of 2BAE, TMDAP, and their blends were exhibited in Figure 3. As can be seen in this figure, absorption rate of TMDAP 30 wt% solution is very slow, it took almost 178 min to complete 90% absorption, while 2BAE 30 wt% solution needed only 61 min. 2BAE5wt%/TMDAP25wt% blend required 127 min and 2BAE10wt%/TMDAP20wt% blend took 109 min to complete 90% absorption. This result indicated that replacing 5 wt% of TMDAP with 2BAE in blend helps reduced reaction time considerably. Similarly, blending with 2BAE helps improve CO_2 absorption kinetics of both 2DMAE and MDEA (Figure 6 and 7). Blending 5 wt% of 2BAE with MDEA do not significantly improve the absorption kinetics compared with single MDEA, however, the same blend concentration of 2BAE with TMDAP and 2DMAE results in considerable improvement of absorption kinetics in comparison with single counterpart tertiary amines. This further confirms that 5 wt% of 2BAE blending with TMDAP or 2DMAE was the best combination, which produced the best amine blend in terms of CO_2 loading, heat of absorption, and absorption rate.

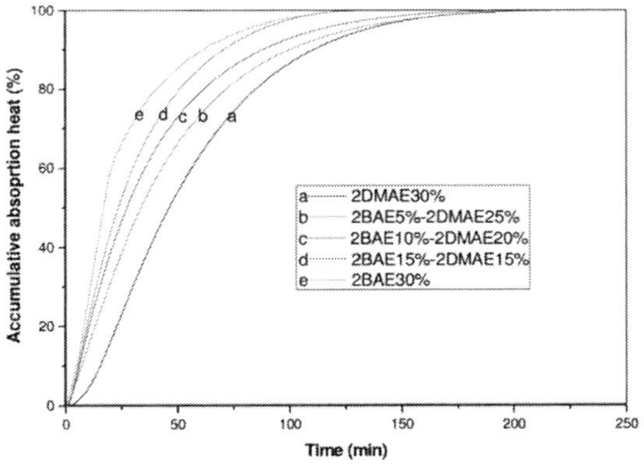

Figure 6. Absorption kinetics of 2BAE/2DMAE blend

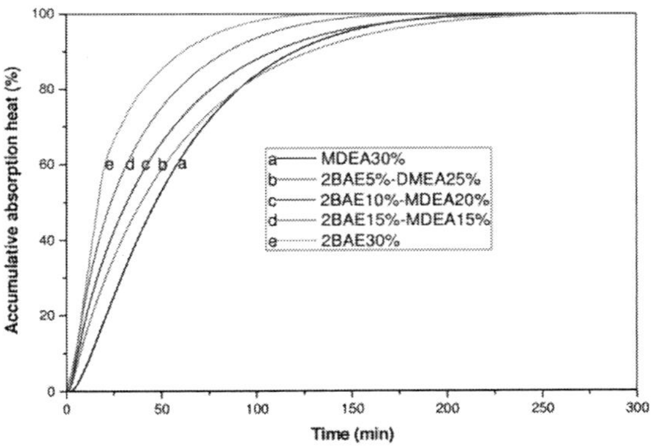

Figure 7. Absorption kinetics of 2BAE/2DMEA blend

Conclusion

Various blends of 2BAE (secondary amine) with three tertiary amines including TMDAP, 2DMAE, and MDEA were prepared and tested. Blending 2BAE with tertiary amine helps improve CO_2 absorption kinetics of tertiary amine, while remains reasonable CO_2 loading and absorption heat. Blends contain 5 wt% of 2BAE provide good CO_2 absorption performance in terms of CO_2 loading, heat of absorption, and absorption kinetics.

References

[1] J.G.J. Olivier, G. Janssens-Maenhout, J.A.H.W. Peters, Trends in global CO_2 emissions, Netherlands Environmental Assessment Agency, 2012.

[2] http://www.esrl.noaa.gov/gmd/ccgg/trends/#mlocited cited on 20/09/2016

[3] IPCC, Intergovernmental Panel on Climate Change (IPCC) Special Report on Carbon Dioxide Capture and Storage, Cambridge University press, Cambridge, UK, 2005.

[4] B. Lee, D. Kim, J. Cho, S. Park, A comparative study on the carbon dioxide capture power between 30 wt% 2-amino-2-methyl-1-propanol and 30 wt% methyldiethanol amine aqueous solutions, Korean J. Chem. Eng., 26 (2009) 818-823.

[5] M.R.M. Abu-Zahra, L.H.J. Schneiders, J.P.M. Niederer, P.H.M. Feron, G.F. Versteeg, CO_2 capture from power plants: Part I. A parametric study of the technical performance based on monoethanolamine, International journal of greenhouse gas control, 1 (2007) 37-46.

[6] M.R.M. Abu-Zahra, J.P.M. Niederer, P.H.M. Feron, G.F. Versteeg, CO_2 capture from power plants. Part II. A parametric study of the economical performance based on mono-ethanolamine, International journal of greenhouse gas control, 1 (2007) 135-142.

[7] A. Adeosun, N. El Hadri, E. Goetheer, M.R. Abu-Zahra, Absorption of CO_2 by Amine Blends Solution: An Experimental Evaluation, International Journal of Engineering and Science, 3 (2013) 12-23.

[8] P. Feron, M. Abu-Zahra, P. Alix, O. Biede, P. Broutin, H. de Jong, J. Kittel, J. Knudsen, L. Raynal, P. Vilhelmsen, Development of post-combustion capture of CO2 within the CASTOR Integrated Project: Results from the pilot plant operation using MEA, 3th International Conference on Clean Coal Technologies for our Futur, Cagliari, Italy, (2007).

[9] R. Idem, M. Wilson, P. Tontiwachwuthikul, A. Chakma, A. Veawab, A. Aroonwilas, D. Gelowitz, Pilot Plant Studies of the CO2 Capture Performance of Aqueous MEA and Mixed MEA/MDEA Solvents at the University of Regina CO2 Capture Technology Development Plant and the Boundary Dam CO2 Capture Demonstration Plant, Industrial & Engineering Chemistry Research, 45 (2005) 2414-2420.

[10] Y. Artanto, J. Jansen, P. Pearson, T. Do, A. Cottrell, E. Meuleman, P. Feron, Performance of MEA and amine-blends in the CSIRO PCC pilot plant at Loy Yang Power in Australia, Fuel, 101 (2012) 264-275.

[11] M.R. Abu-Zahra, Z. Abbas, P. Singh, P. Feron, Carbon Dioxide Post-Combustion Capture: Solvent Technologies Overview, Status and Future Directions, Materials and processes for energy: communicating current research and technological developments, Formatex Research Center2013, pp. 923-934.

[12] F.-Y. Jou, A.E. Mather, F.D. Otto, The solubility of CO2 in a 30 mass percent monoethanolamine solution, The Canadian Journal of Chemical Engineering, 73 (1995) 140-147.

[13] A. Chakma, CO2 capture processes — Opportunities for improved energy efficiencies, Energy Conversion and Management, 38, Supplement (1997) S51-S56.

[14] N. El Hadri, D.V. Quang, E.L. Goetheer, M.R.A. Zahra, Aqueous amine solution characterization for post-combustion CO 2 capture process, Applied Energy, (2016).

465184 Economic and Exergy Analysis of Integrated Purification Process with Carbon Capture Process (CCP) Using Selexol in a 500 MW Integrated Gasification Combined Cycle (IGCC) Power Plant

Abstract:

The enormous CO_2 emission from power plants raises the significant effects on global climate. In order to minimize greenhouse gas emissions from the utilization of fossils fuel, integrated coal gasification combined cycle (IGCC) with carbon capture process (CCP) takes a spotlight for high efficient and eco-friendly electricity generation. In this study, an integrated purification process with CCP for a 500MW IGCC power plant was simulated to evaluate energy efficiency and cost. First, only dual-stage Selexol processes with 80%, 85%, 90% and 95% carbon capture were preliminary simulated. According to preliminary results, energy consumption of Selexol process showed an optimum at more or less 90% carbon capture. Since each unit is integrated with the other units, the operating condition of one can affect the others. Then, as a second step, an integrated purification process with CCP simulation was carried out to evaluate energy analysis and cost estimation. It consists of contained water gas shift reactor (WGSR), dual-stage Selexol process, Claus process, scrubber unit, sour water stripper process, hydrogenation process, drying process, and compression. Finally, the exergy sankey diagram for whole integrated purification process with CCP, based on second law of thermodynamic, was performed to improve overall capture efficiency. Then, the development direction for reducing energy consumption was suggested from the exergy analysis.

470573 Influence of Alloying on CO_2 Electroreduction on Ag-Zn System

Abstract:

Electrochemical reduction of CO_2 is an attractive pathway for the sustainable production of fuels and chemicals, as it allows for the effective use of renewable energy sources while mitigating the effects of CO_2 in the atmosphere. While significant decrease in overpotentials and increase in reaction rates have been reported for the two-electron reduction products, reports on improved activity and selectivity for the products that require more than two electrons (> 2 e^- products) have been limited on monometallic transition metal catalysts. Recent theoretical work indicates that this difficulty of reducing CO_2 to > 2 e^- products is due to the linear scaling of the binding energies of CO_2 reduction intermediates on transition metal catalysts, and the necessity of catalyst systems that can break such scaling has been emphasized.(1)

Alloys, which can provide bifunctional sites that selectively stabilize intermediates, are a promising class of materials, and a large amount of effort has been focused on the investigation of their effectiveness as CO_2 reduction catalysts. Recent works have reported on improved CO_2 reduction activity and selectivity on alloy systems, which were attributed to the synergy between the two metal atoms at the surface. However, most improvements reported in these works are pertaining to two-electron products, and alloy systems that demonstrate improvements for > 2 e^- products remain elusive.

Here, we report on the investigation of polycrystalline AgZn foil, which resulted in the observation of enhanced activities and selectivities for methane and methanol productions compared to Ag and Zn foils. Although CO is the main CO_2 reduction product as expected of an alloy between two metals active for CO production, the selective enhancement observed for the > 2 e^- products is noteworthy, as it indicates that alloys are promising candidates in the development of effective CO_2 electroreduction catalysts. The production of CO as major product is explained by the binding energies of intermediates, and the importance of optimal binding energies are discussed.

References

1. A. A. Peterson and J. K. Nørskov, *The Journal of Physical Chemistry Letters*, **3**, 251 (2012).

470038 *In-Situ* Infrared Spectroscopic Investigation of Pyridine-Mediated CO_2 Reduction on Pt Electrocatalysts

Abstract:

The pyridine-mediated electrochemical reduction of CO_2 to methanol, formate, and formaldehyde was first demonstrated by Bocarsly et. al in 1994. Since then, many mechanistic investigations of this process have been conducted using reactivity, cyclic voltammetry, or computational methods; however, direct experimental mechanistic investigation has been lacking. We employ surface-enhanced infrared absorption spectroscopy, coupled with traditional electrochemical techniques in order to probe the electrochemical interface during the pyridine-mediated electrochemical reduction of CO_2 on Pt electrocatalysts. Previously proposed mechanisms are discussed and refined in light of new insights obtained in these studies.

476849 Next-Generation High-Efficiency Hybrid Solar to Hydrogen Conversion with Integrated Storage

Abstract:

By 2050 solar technologies are projected to deliver the majority of the world's electricity. Solar energy can be used to generate both heat and electrical power. Most solar panels are designed for only one of these purposes, so an electrical photovoltaic (PV) panel is typically less than 20% efficient.

It is well known that PV cells experience a deterioration in performance (efficiency) when they are operated at high temperatures, and that this leads to particularly high losses when the solar resource is at its highest. For example, a drop in PV cell efficiency in the range 10-20% can be expected, depending on the PV technology, when operating PV cells at temperatures of ~65 °C (or, 40 °C higher than room temperature), which is easily experienced in hot climates. Hybrid PV-thermal (PV/T) solar collector technology combines PV modules with the contacting flow of a cooling fluid (gas or liquid) in a number of configurations and geometries, and offers some advantages when space is at a premium and there is demand for both heat and power. Moreover, a major motivation in PV/T technology is to cool the PV cells and therefore to increase their electrical efficiency, while delivering a potentially useful thermal output (hot fluid stream). PV/T collectors are a highly efficient technology, capable of achieving system efficiencies (electrical plus thermal) in excess of 70%.

By far the most common use of the thermal-energy output from PV/T systems (in fact most solar-thermal collector technologies) is to provide hot water at 50-60 °C for households or commercial use, however, a wide range of opportunities arise at higher temperatures whence additional power-generation cycles (e.g., with organic Rankine cycles, thermoelectric generators, amongst other) or thermal-cooling technologies (e.g., with desiccant, ad/absorption refrigeration cycles, amongst other) can be combined with PV/T modules. These additional options become viable at temperatures typically above ~80 °C, and importantly, become increasingly efficient at progressively higher temperatures.

Operating PV/T panels efficiently at high temperatures for this purpose is a significant scientific and engineering challenge, since the two modes of heat-transfer loss that result directly in the drop in performance, namely convection and radiation, are both exacerbated at higher temperatures. The reduction of convective losses can be addressed by evacuation of the panel, while radiation losses can be decreased by using selective coatings. This talk will discuss recent advances in evacuated PV/T technology with optimized surfaces and selective coatings combined with a secondary power generation sub-system based on organic Rankine cycles (ORCs) for high-efficiency conversion of solar energy to electrical power and consequently hydrogen via electrolysis. Projections at this stage suggest an improvement of the order of 20-30% relative to PV-only systems, for the same area.

In this solar-to-electricity-to-hydrogen system, energy storage options include thermal storage, electrical storage and hydrogen storage. Of these, thermal storage is relatively low cost and can be performed over a range of scales with sensible heat storage (e.g., hot water tanks), phase-change materials (PCMs) covering a range of organic or inorganic substances (e.g., molten salt tanks), or via thermo-chemical energy storage which utilizes the energy alternately absorbed and released in reversible chemical reactions and has the highest theoretical storage capacity of these solutions. Electrical-energy storage on the other hand is highly dependent on scale. At small scales of application a variety of battery technologies exist or are currently under development, which are highly suitable solutions. At large-scales, on the other hand, the costs and potential environmental impact of the widespread use of batteries (arising from the use of hazardous or scarce substances) allow serious competition from mechanical or thermo-mechanical systems, which become the technology of choice. Relevant conventional electricity-storage solutions include pumped hydro (PHS) and compressed air energy storage (CAES).

A significant effort has been placed in recent years on the development of additional electricity storage technologies, driven by the increasing penetration into the grid of renewable energy technologies for electricity generation, many of which have an unpredictably intermittent nature. Of increasing interest are the so-called cryogenic 'liquid-air' (or, liquid-nitrogen) energy storage (LAES) system and a novel storage technology known as Pumped Thermal Electricity Storage (PTES). LAES uses excess electricity to operate a refrigeration cycle with which to liquefy air, which is then stored in a tank at atmospheric pressure. When the electricity is needed, the liquid air is evaporated and used to drive a turbine. PTES, on the other hand, uses excess electricity to operate a heat pump cycle with which to convert the electrical energy into thermal energy. The thermal energy, both hot and cold relative to atmosphere, is then stored in two large reservoirs (hot and cold). The reservoirs contain a material with a high heat capacity, and are able to store the energy much more compactly than PHS. When required, the thermal energy can be converted back into electricity by running the heat pump in reverse, as a heat engine. The projected round-trip efficiencies of both LAES and PTES are lower than PHS, but these systems have a number of potential benefits, including low capital cost, no geographical constraints (i.e., elevations, caverns, etc.) in terms of deployment, and no associated environmental impact. This talk will present recent advances in the development of these technologies, compare the various options with each other and with alternative thermal and hydrogen storage in the context of solar-energy conversion and storage, and finally discuss the next-steps that would enable their successful deployment. In isolation, thermal-energy storage emerges as the technological option with the lowest cost and highest technical maturity, whereas at larger scales electricity storage can benefit as part of a wider storage, transmission and distribution network that includes additional intermittent generation from renewables.

481381 Understanding and Eliminating Interaction of Cells in an Electrochemical Stack

Abstract:

Typically, both the fuel cells and electrolysis cells are stacked in series by means of impermeable graphite or metallic bipolar plates. The open-circuit voltage of most types of fuel cells is approximately 1 V. To support sufficient voltage and power required for commercial applications, cells are assembled in a stack. By connecting multiple cells in series, the voltage of a stack is multiplied, generating the required high power output. Similarly, electrolysis cells are stacked to uptake the high voltages and power, and reach a demanding hydrogen production; the hydrogen generation-rate of a single cell is limited to electrodesâ€™ active area.

In an ideal stack, the design and operating conditions for all cells should be identical throughout the stack. Overall stack performance would then be the summation of the equal stand-alone performances of cells. However, in a real stack, the cells are not perfectly identical and they might differ either in design or operating conditions or both for a variety of reasons: imperfections during manufacturing, a non-uniform inlet flow distribution, degradation of one or more cells and external boundary conditions, such as heat transfer with the surroundings. Though such non-uniformities are encountered at a single cell level, the performance differences are translated, very often not to a single cell only, but to a group of cells located in neighborhood. In other words, one poorly operating or perturbed cell could affect the performance of other cells in the stack. Such a coupling of the cells is the focus of investigation here as it could be detrimental for the performance and lifetime of the stack.

As described above, the individual cells in a stack are separated by impermeable bipolar plates. Thus, we can classify the transport phenomena in their stacks according to the region of influence. Mass, momentum and species transport are confined to individual cells, whereas charge and heat transfer take place globally across the cells such that the cells become coupled and influence each other. Since the cells are coupled via heat and charge transport, abnormal behavior of one cell can propagate to the neighboring cells. Two adjacent cells can be said to be (1) electrically decoupled when their local current density distributions do not influence and perturb each other, and (2) thermally decoupled if there is negligible heat transfer between the two cells. We analyze charge and heat transport across the cells in a stack for the above two decoupling conditions. In short, electrical decoupling of the cells in a stack is found to depend on the geometrical and material properties of the bipolar plate and the operating parameters. Similarly, the thermal decoupling is governed by the design and operating parameters of coolant flow fields. We finally identify the design guidelines that ensure decoupling of the cells via both the transport modes. The guidelines could be of use for stack designers when selecting the bipolar plate material and dimensions and coolant flow field conditions, so as to minimize the spread of disturbances in a cell to neighboring cells.

Acknowledgements: The research is supported by the National Research Foundation, Singapore under its Competitive Research Program (Award No. NRF-CRP8-2011-04).

454078 Advancement of Hydrogen Storage Technologies

Abstract:

Hydrogen is seen as one of the major storage options for renewable energy. However, its storage still faces a number of challenges. Safe, dense and efficient hydrogen storage has to overcome a number of thermodynamic constraints. A big issue, for example, is the compromise between simple hydrogen release with a small energy demand on the one side and loss-free storage at ambient conditions with a high density on the other side. The interplay between enthalpy and entropy change during hydrogen uptake plays an important role in this respect.

In this contribution different steps in the development of hydrogen storage technologies, such as Liquid Organic Hydrogen Carrier (LOHC), will be presented. First, a study on the identification of potentials for the improvement of hydrogen storage technologies is conducted. This outlines the impact of different measures and gives insight into the question what should be improved.

Based on these findings, a systematic development of carrier materials and technologies will be demonstrated. A comparison of different hydrogen storage options is discussed.

The role of process simulation in this context will be outlined. In the last step a special emphasis will be put on the issue of reliability and uncertainty of the respective process models.

458291 Thermodynamic and Economic Evaluation of Hydrogen Transport Technologies

Abstract:

Due to its high gravimetric energy density, hydrogen is considered as an interesting energy carrying substance for a future energy system based on renewable resources. In this context, energy distribution and thereby efficient, cheap and secure transport of hydrogen is an important aspect.

Today, hydrogen transport is almost exclusively realized by compressed or liquefied gas. Transportation as compressed gas affords high pressures, generating a certain security risk while the amount of transported hydrogen is still low due to the low storage density. Liquefied gas can reach substantially higher storage densities and thereby capacities but suffers from large energy demands for liquefaction and boil off-losses during storage. Several alternative storage technologies came into the focus of research to overcome these limitations. In the context of hydrogen transport, especially carrier-based hydrogen storage technologies like metal hydrides or liquid organic hydrogen carriers (LOHCs) seem promising as these technologies enable a dense storage of hydrogen at ambient conditions.

In this contribution, a thermodynamic and economic evaluation of several hydrogen transport technologies is presented. The analysis commences with the determination of the energy demands for hydrogen conditioning (e.g. compression, liquefaction), the actual transport and the provision of hydrogen gas at the point of use (e.g. hydrogen liberation from carrier). For all technologies considered in the analysis the energy demand for the actual transport is comparatively low. The conditioning (for pressurized and liquefied gas) and provision step (for carrier-based transport) have a far higher energy demand. Options to reduce the energy demands are pointed out. The cost calculation includes investment, maintenance, staff, energy, and material costs and shows the potential advantage of chemical carrier technologies over the established transport technologies.

459518 State-of-the-Art of the Liquid Organic Hydrogen Carrier (LOHC) Technology

Abstract:

LOHC are a safe and dense solution of storing hydrogen at ambient conditions. The basis is a reversible chemical reaction of hydrogen with the LOHC. This technology is studied worldwide since 5 years.

Despite the fact that several component pairs (loaded and unloaded form) are able to be a LOHC, the basic process design is known.

The lecture highlights technical applications of this technology. This progress will push the development of a hydrogen-based energy system.

481365 On using CO_2 and Renewable Energy for Autonomous Cities

Abstract:

An innovative concept of district heating using carbon dioxide is proposed to supply energy services in cities. The principle is to use CO2 as a district heating fluid distributed in liquid and gas form at a temperature of 15-20 °C. Heating needs are then supplied by condensing CO2 while cooling needs are supplied by evaporation, the temperature levels being adjusted using heat pumps. The energy balance is closed by exchanging heat between the district heating fluid and the environment accessing therefore attractive sources. The advantage of such system is the supply of the heating/cooling services at the temperature it requires and the possible exchange between heating and cooling demands. The use of the district heating system allows indeed to connect the most attractive environmental sources (geothermal, lake, river, waste water, industrial waste heat,. . .) with the users in order to maximise the efficiency of the system. Using this type of technology, it has been demonstrated that heating and cooling the city center of a European city can be reduced by a factor 5 to 6 (i.e. 80% energy savings) while having a pay back period of around 7 years . The concept can be extended to integrate the renewable energy resources in particular solar photovoltaics (PV). One of the difficulty of the PV systems is the excess of electricity available in the summer that needs to be stored to supply services in the winter. The idea here is to electro-chemically convert excess of electricity into methane by combining electrolysis of water and the Sabatier reaction that converts CO2 and H2 into methane. This can be done in co-electrolysis concepts
using solid oxide electrochemical membranes. Methane can then be stored in liquid form. In the winter time, the methane is converted in decentralised solid oxide fuel cells into electricity and heat while separating the CO2. The CO2 is collected in the district CO2 network at 50 bar and stored in the liquid form for the summer season when it will again be converted into methane. The produced electricity is used to supply the needs of the households, to drive the heat-pumps of the CO2 based district network and even electrical cars. It is therefore possible to develop cities that are supplied only by renewable sources. Furthermore, the methanation concept can also be applied to convert the waste biomass produced in the city into methane.

459469 Nano-Sized Nioâ€"SDC Composite Synthesized Via a Low-Temperature Hydrothermal Route in a Single Step As the Anodes of Intermediate Temperature Solid Oxide Fuel Cell

Abstract:

A solid oxide fuel cell (SOFC) is considered as a high efficient energy converting device for chemical energy in fuels directly to electrical energy via an electrochemical reaction, getting rid of the efficiency limit of the Carnot cycle and with minimal emission [1]. In the last two decades, a promising composite electrolyte material composed of a doped ceria phase and a binary carbonate salt phase has been tested due to its efficient ionic conductivity and promising performance in intermediate temperature SOFC [2-5]. However, the anode material of the composite electrolyte-supported SOFC has still been a challenge. In this work, homogenous nano-sized NiO-samarium doped ceria (marked as HT-NiO-SDC) composite with uniform particle size and a good sintering ability was synthesized through a low-temperature hydrothermal route, and used as the anode of SOFC with a composite electrolyte. For comparison, the conventional solid mixing sample (denoted as SM-NiO-SDC) was also prepared. The electrical performance of the single cells with the NiO-SDC anode and SDC-$(Li_{0.67}Na_{0.33})_2CO_3$ composite electrolyte was investigated.

The cubic NiO phase and fluorite-structured SDC phase in the HT-NiO-SDC anode were confirmed with XRD. The diameter of the HT-NiO-SDC particle observed with SEM and TEM is around 10 nm. The electronic conductivity of HT-NiOâ€"SDC cermet measured with a DC four-probe technique in H_2 is much higher than that of the conventional SM-NiO-SDC sample. Moreover, the electrochemical impedance spectroscopy result shows that the polarization resistance of HT-NiO-SDC is much lower than that of SM-NiO-SDC, indicating a homogeneous distribution of the NiO and the SDC phases in the HT-NiO-SDC sample.

Fig. 1a shows that the performance of the cell with HT-NiO-SDC anode improves with the increase of the NiO

content in the anode. The maximum power density of the cell is 738 mW cm^{-2} at 700 °C, much higher than that of the cell with conventional SM-NiO-SDC anode, as shown in Fig. 1b, demonstrating that the hydrothermal synthesis is a promising method to preparing anode materials for intermediate-temperature SOFCs with a ceramic-carbonate composite electrolyte.

462429 Activity and Selectivity Trends on Bimetallic Thin Films for Electrochemical Carbon Dioxide Reduction

Abstract:

Utilizing carbon dioxide (CO_2) as a feedstock for fuels and industrial chemicals is ideal owing to its natural abundance and the desire to reduce atmospheric concentrations of this greenhouse gas. The ability to electrochemically reduce CO_2 to these value added products has been demonstrated under moderate conditions (ambient temperature, pressure and in aqueous solutions) [1], yet the current state of electrocatalyst technology is not sufficiently advanced to the point that CO_2 reduction devices have become technologically relevant. At the present time, copper is the only transition metal shown capable of producing high proportions of attractive products, including methane, ethylene and ethanol [2]; albeit at very high overpotentials and with limited product tunability. It is of great importance to overcome these current limitations and develop electrocatalysts with high activity and good selectivity towards practical fuels and industrial chemicals.

In this work, we have developed bimetallic electrocatalyst thin films by electron beam evaporation. Their activity and selectivity are investigated towards CO_2 electroreduction in our three-compartment electrochemical cell that allows for excellent liquid and gaseous product detection [3]. We target certain bimetallic arrangements, including phase segregated systems, overlayers and near surface alloys. These configurations allow us to capitalize on ensemble and/or ligand effects at the electrocatalyst surface [4] with the goal of targeting further reduced (> 2 electron) CO_2 reduction products at high Faradaic efficiencies. By correlating thin film characterization before and after testing with results of electrochemical measurements, we draw surface structure-activity trends among this class of materials. Detailed understanding of these trends can be leveraged by the electrocatalysis community to move beyond pure copper based systems for electrochemical CO_2 reduction.

References

[1] Hori, Y., *Modern Aspects of Electrochemistry*, 42 (2008) Springer, New York, 89-189.

[2] Kuhl, K., Hatsukade, T., Cave, E., Abram, D., Kibsgaard, J., Jaramillo, T., *Journal of the American Chemical Society*, 40 (2014) 14107-14113.

[3] Kuhl, K., Cave, E., Abram, D., Jaramillo, T., *Energy & Environmental Science*, 5 (2012) 7050-7059.

[4] Rodriguez, J., Goodman, W., *Journal of Physical Chemistry*, 95 (1991) 4196-4206.

465569 Asymptotically Reduced Zero-Dimensional Model for an All-Vanadium Redox Flow Battery

Abstract:

Recently, mathematical modeling and numerical simulations have come to play an important part in the research and development of all-vanadium redox flow batteries (VRFBs). Detailed mechanistic models consider conservation of mass, momentum, species, energy and charge transfer in the various layers (current collectors, electrodes, and membrane) and thus result in a computationally intensive system of coupled partial differential equations (PDEs). The associated computational cost for a system of PDEs typically decreases as the number of spatial dimensions is decreased. Hence, the lower dimensional models would be more suitable for wide-ranging parametric studies of a single cell, stack or system, real-time optimization and feedback control. There is thus a need to derive reduced dimensional models that can predict the cell behavior at a minimal computational cost while preserving the essential physics.

In this work, we present a reduced zero-dimensional model that both preserves essential physics and reduces computing time. The methodology uses scale analysis and asymptotic reduction. The model reduction from 3D to 2D is straightforward and does not require detailed analysis. The porous nature of the carbon felts and the membrane and the solid nature of the current collector allow for a reduction from three to two dimensions because changes in the dependent variables in the spanwise direction are negligible; hence a 2D model ought to be able to capture the behavior of a 3D cell. However, further reduction in the dimensionality, i.e., 2D to 1D or 0D needs to be carefully analyzed. In this regard, we carry out scale analysis of a 2D model accounting for conservation of species and charge in a VRFB to derive a reduced 0D model. The conditions when a 0D model can capture the behavior of a 2D cell are identified in terms of two non-dimensional numbers and their limits. We verify the reduced model by comparing its charge-discharge curve predictions with that of a full two-dimensional model. The reduced model takes the form of an explicit algebraic equation and can be implemented in a high level programming language such as Fortran or C with a minimum computation overhead, i.e., only few kilobytes of memory and on the order of milliseconds for the complete charge-discharge curve. The proposed analysis leading to reduction in dimensionality is generic and can be employed for other types of redox flow batteries.

474127 Formic Acid Based Chemical Hydrogen Storage for Fuel Cells

Abstract:

The success of the "hydrogen economy" will ultimately depend on efficient and sustainable production, storage, and distribution of hydrogen [1-2]. Owing to its low toxicity and high volumetric H_2 storage capacity, formic acid (FA) has been considered as one of the promising chemical hydrogen storage materials for fuel cells and energy storage systems. However, for FA to be useful for practical applications, highy active catalysts for FA dehydrogenation are necessary. In this contribution, we present highly capable Pd-based nanoparticles supported on nitrogen doped carbon supports including Pd/C_3N_4 [3], Pd/SBA-15-Amine [4], and a surface-modified PdAu nanocatalysts. We further discuss about the potential influence of the added nitrogen eleement on FA dehydrogenation activities.

References

[1] J. O. M. Bockris, *Science* **176** (1972), 1323.

[2] J. Yang, A. Sudik, C. Wolverton, D. J. Siegel, *Chem. Soc. Rev.* **39** (2010) 656–675.

[3] J. H. Lee, J. Ryu, J. Y. Kim, S. W. Nam, J. H. Han, T. H. Lim, S. G., K. H. Chae, C. W. Yoon, *J. Mater. Chem. A* **2** (2014), 9490-9495.

[4] K. Koh, J.-E. Seo, J. H. Lee, A. Goswami, C. W. Yoon, T. Asefa, *J. Mater. Chem. A* **2** (2014), 20444-20449.

474207 *Ab Initio* modeling of the Cation Diffusion in $La_{1-x}Sr_xMnO_{3\pm\delta}$ for Solid Oxide Cell Electrodes Applications

Abstract:

Cation diffusion in $La_{1-x}Sr_xMnO_{3\pm\delta}$ (LSM) and in related perovskite materials play an important role in controlling performance and long term stability of solid oxide fuel cells and electrolytic cells. In particular, cation transport contributes to the formation of the secondary phases, segregated surface compositions, and diffuse interfaces between cell components. Fundamental mechanistic understanding and quantitative theoretical model predictions on cation diffusivity in LSM are still insufficient to accurately reproduce experimental cation diffusivity measurements. In this work, based on *ab initio* based LSM defect modelling [1] we determined the defect interaction energies for defect clusters involved in cation diffusion, and the migration barriers of candidate diffusion mechanisms. This data has been further used to calculate the $La_{0.75}Sr_{0.25}MnO_3$ cation self-diffusion coefficients vs. 1/T (T=873~1273 K) from a random-walk diffusion model. The predicted Mn self-diffusion coefficients, along with those of La/Sr cation diffusion [2], will be discussed and compared with experimental tracer diffusion coefficients reported in the literature [3].

[1] Y.-L. Lee and D. Morgan, Phys. Chem. Chem. Phys., 2012, 14, 290–30

[2] B. Puchala, Y.-L. Lee, and D. Morgan, J. Electrochem. Soc., 160 (8) F877-F882 (2013)

[3] S. P. Harvey, R. A. De Souza, and M. Martin, Energy Environ. Sci., 2012, 5, 5803

474554 Microwave Assisted Hydrogen Production from Ammonia over Carbon Supported Materials

Abstract:

In this study, hydrogen production from ammonia was investigated in a focused microwave heated system and the results were compared with the corresponding results obtained in a conventionally heated tubular reactor.

While fossil fuels have been still used as a main energy resource, alternative energy sources have gained great attention due to the decrease in the fossil fuel reserves and the increase in the concentration of CO_2, which is emitted to the atmosphere due to the usage of fossil fuels. Utilization of hydrogen as an alternative energy source especially by means of fuel cell application increases steadily, due to its energy content and its clean fuel properties. Since, the purity of hydrogen is very important especially for PEM fuel cells, in this study ammonia has been selected as a raw material for hydrogen production since it has high hydrogen content and can be stored and distributed easily by means of conventional methods as well as COx free hydrogen can be produced by decomposition of ammonia.

Nowadays, microwave energy is used as an alternative heating method to enhance heterogeneous chemical reactions, especially for endothermic ones. Differently from conventional methods, such as electrical resistance furnace, direct conversion of electromagnetic energy into heat can be achieved through microwave system instead of heat transfer. Therefore a temperature distribution in radial direction within the catalytic bed is avoided in comparison to conventionally heated system.

In this study, microwave assisted ammonia decomposition reaction was carried out over transition metal incorporated carbon supported catalysts. Carbon materials were used due to its superior properties as microwave absorbers in the reaction medium. Catalysts were synthesized following the impregnation procedure at different metal loadings and they were characterized by different techniques, namely, TGA, TPR, XRD, nitrogen physisorption, ICP, HRTEM/EDX

after calcination, reduction and reaction steps. When the performances of the synthesized catalysts were evaluated in the conventional system with a GHSV of 36,000 ml/hg_{cat}, negligible activity was observed below 500°C. In the case of microwave system, conversion of ammonia was observed at temperatures as low as 300°C. Moreover complete conversion was achieved over the synthesized catalysts at a reaction temperature between 400-500°C, such as, for mesoporous carbon supported molybdenum, total conversion was found at 400°C.

Acknowledgement: Financial support of TUBITAK 214M148 was gratefully acknowledged.

476635 Three-dimensional electrode for Electrodialysis unit

Abstract:

Surging population and climate change will push society, ever more urgently, to harvest lower quality or impaired water supplies (e.g., sea water or brackish groundwater) for drinking water. For example, desalination of sea water has become a promising technique for providing additional sources of fresh water along coastal communities. However, the widespread adoption of desalination technologies (Reverse Osmosis and electrodialysis) is currently limited due to its energy intensive nature and/or limited membrane lifetime and performance, especially during high operation rates. This work focuses on developing a novel, cost-effective, integrated sunlight driven electrodialysis (ED) unit. Specifically, we have replaced planar electrodes in traditional electrodialysis units with 3-D circulating photoelectrodes that generates the required electrochemical potential gradient upon illumination to drive ion transport across the membranes. The 3-D circulating photoelectrode bed configuration provided large surface area per unit volume, thereby allowing high currents to be drawn at current densities short of the diffusion-limited regime, a criteria needed to prevent unwanted secondary reactions such as water splitting. Effect of temperature (20, 30, and 40 â°C) and flow rate of sea water (0.1, 0.2, 0.4 ml/s) on salt-ion rejection was also investigated.

455098 The Sunshot Inititiative: New Opportunities for R&D in Concentrating Solar Power

Abstract:

The SunShot Initiative is a collaborative, national endeavor by the United States Department of Energy's Solar Energy Technologies Office to make electricity generated from the sun cost competitive – on an unsubsidized basis – with traditional forms of electricity generation by the year 2020. Concentrating Solar Power (CSP) constitutes one technological approach to achieving this ambitious goal, generally identified as a levelized cost of electricity (LCOE) ≤ $0.06/kWh_e$. CSP focuses the sun's rays onto a receiver that converts the photons into thermal energy. Part of this thermal energy can then be transported to storage for use when the sun does not shine, while the remainder is transported to a conventional turbine to generate electricity. The turbine, or heat engine, can realize increased thermodynamic efficiency as the input temperatures increase; however, high temperature CSP systems come with significant materials and cost challenges. Evaluating the interfaces found in the solar receiver, thermal energy storage technologies, heat exchangers, and power blocks will help define the technical challenges and their requisite success metrics. The presentation concludes with remarks, links, and information on DOE's SunShot Initiative so that the audience may stay aware of future Funding Opportunity Announcements.

469740 Non-Stoichiometric Perovskite Oxides As High-Temperature Energy Storage Media and Their Application to Concentrating Solar Power Generation and Hydrogen Production

Abstract:

Although there is a significant abundance of solar resource, the intermittency of the resource compromises its deep integration into the U.S. energy portfolio. However, solar-thermal processes such as concentrating solar power (CSP) are unique in that thermal energy can be efficiently stored in engineered heat-absorbing media. The quantity of heat stored and subsequent power-generation efficiency generally increase as temperature increases, motivating solid particulate storage media for high-temperature systems. However, the storage of only sensible energy limits the potential of both the quantity of stored energy as well as the possible applications of the stored energy. For this reason, energy storage media have been engineered to store both sensible heat and reaction heat, a practice known as thermochemical energy storage (TCES).

Here we present the concepts behind the engineering, synthesis, and characterization of novel doped calcium manganese perovskites ($CaAl_yMn_{1-y}O_{3-\delta}$). These materials exhibit rapid and reversible reduction/oxidation behavior through the introduction of oxygen vacancies in the material structure: $ABO_3 + \Delta H \rightarrow ABO_{3-\delta} + \delta/2\ O_2$. This approach allows the storage of chemical energy through thermal reduction without the kinetic limitations inherent in decomposition reactions. The dopant concentrations of this materials family have been optimized in our laboratories to produce the largest energy storage capacity of a high-temperature perovskite system to date.

Additionally, two applications of the material are presented. First, these materials are applied to a CSP system for electricity generation. In this system, the material is first heated and thermally reduced in the incident solar beam of a solar receiver. The material is then stored in large thermally-insulated bins, after which both the sensible and chemical energy can be utilized by a reaction and heat-transfer process with a working fluid. This high-enthalpy working fluid is passed to a turbine to create power on demand, even when the solar resource is unavailable. Secondly, these materials can operate as thermal-to-chemical energy converters, where the thermal reduction results in a chemically energy-rich species which can be used either fully or in part to facilitate subsequent reactions, such as water-splitting for hydrogen generation. In this embodiment, only the chemical energy is useful, and the sensible energy must be recovered to maximize process efficiencies.

Sandia National Laboratories is a multi-program laboratory managed and operated by Sandia Corporation, a wholly owned subsidiary of Lockheed Martin Corporation, for the U.S. Department of Energy's National Nuclear Security Administration under contract DE-AC04-94AL85000. This work is supported by the U.S. Department of Energy, SunShot Initiative, under Award Number DE-FOA-0000805.

471159 Solar Thermochemical Energy Storage Based on Strontium Carbonate Chemistry

Abstract:

The $SrO/SrCO_3$ chemistry offers the potential for high temperature heat release (~1100-1200°C) upon the carbonation of SrO, which could be a pathway to high efficiency, combined cycle power generation. Building upon previous work on this cycle, this talk will propose a kinetic model for the carbonation reaction derived on experiments using both TGA and packed bed systems. In addition, preliminary technoeconomic and exergy analyses are performed for the cases of CO_2 storage as 1) a compressed gas and 2) ab/adsorbed species in a liquid or solid matrix. Probabilistic methods are used to elucidate the cost drivers and areas for future consideration in evaluation of the cycle feasibility.

452449 Design of Manganese Oxide-Based Particles for High-Temperature Thermochemical Energy Storage

Abstract:

High-temperature thermochemical energy storage shows promise in aiding concentrating solar power plants in meeting variable, grid-scale electricity demand. In this work, manganese oxide-based mixed metal oxide particles have been designed and tested for thermochemical energy storage. Particles are designed for high energy storage capacity, flowability, and physical and chemical stability. We evaluate the effects of Al_2O_3, Fe_2O_3, and ZrO_2 in Mn_2O_3-based spray-dried particles in a TGA between 650°C and 1,200°C over six consecutive redox cycles. Results are compared with thermodynamic predictions from 400-1,400°C under oxidizing and reducing atmospheres. A mixture of 2:1 Fe_2O_3:Mn_2O_3 formed iron manganese oxide spinel (Fe_2MnO_4) on calcination, and demonstrated the highest thermochemical activity despite particle agglomeration and deformation. Conversely, zirconia was an inert support that does not react with manganese oxide. Differences in redox performance between materials with different Fe to Mn ratios have been attributed to ion diffusion and secondary phase formation.

452774 Grid-Scale Thermochemical Energy Storage Using Mixed Metal Oxide Redox Cycles

Abstract:

Thermochemical energy storage (TCES) systems are of great interest in concentrated solar power (CSP) applications. Storing sunlight as chemical energy during the day can enable power generation at night or during cloudy periods, effectively alleviating the inherent intermittency of solar sourced electricity. Metal oxides are among the most attractive TCES materials because they possess high-energy density and high reduction/oxidation temperatures (>1000°C) suitable for driving high-efficiency thermodynamic power cycles. However, limitations involving high-temperature heat transfer, particle handling, sintering, and chemical performance of the metal oxide materials are still problematic in practical application.

A novel, grid-scale CSP/TCES system design that addresses many of the traditional problems associated with metal oxide TCES has been developed at the ANU. This presentation serves as a broad overview of recent work surrounding the designed CSP process, including improvements to the metal oxide active materials, solar reactor design, optical field development, and preliminary techno-economic analyses. Improved chemical performance of the active material is being explored through metal co-doping and the intentional formation of solid solutions. The incorporation of fluidized bed reactors for solar reduction and off-sun oxidation allows for improved heat transfer and enhanced chemical kinetics, and also enables high temperature pneumatic transport of gas-solid reacting flows. The initial solar reactor concept consists of one or several vertical fluidized bed reaction tubes in a beam-up solar cavity receiver, which allows for well-controlled fluidization, high optical efficiency, and minimized convective heat losses. The techno-economic analysis of a 100 MW_{Th} grid-scale facility is employed to predict the levelized cost of electricity.

465618 Kinetic Analysis of High Temperature Thermochemical Energy Storage Based on Calcination–Carbonation Chemical Looping Reactions

Abstract:

This work aims to improve the performance of carbonate storage systems, a form of thermochemical energy storage (TCES) for concentrated solar power (CSP) technologies. TCES has great potential due to its high energy storage densities and inexpensive materials, making it highly competitive with commercial molten salt storage systems. The storage system is based on metal oxide calcination–carbonation chemical looping, a cyclic process for which a broad range of metal oxide candidate materials are considered. Metal oxide produced in the solar-driven endothermic calcination reaction is cycled to the exothermic carbonation reaction, from which thermal energy is released to drive a power cycle.

This work details a comparative thermodynamic and kinetic study of calcination and carbonation reactions, using commercial samples and synthesized mixtures of carbonates such as $CaCO_3$ and $SrCO_3$. Experiments are carried out under a CO_2 atmosphere using thermogravimetric analysis (TGA) and differential scanning calorimetry (DSC) and the materials are selected to attain the desired storage temperature (typically above 565°C), energy density and charging and discharging rates. Chemical and structural cyclability of candidate materials are evaluated as a function of the calcination–carbonation cycle number using several characterization techniques such as scanning electron microscopy (SEM), transmission electron microscopy (TEM), N_2 adsorption-desorption and X-ray diffraction crystallography. We show how different mixtures and synthesis precursors can improve the cyclability of metal oxide–carbonate systems.

465613 Thermodynamic Analysis of $CaCO_3$/CaO Looping for High Temperature Thermochemical Energy Storage

Abstract:

High temperature power cycles, such as the supercritical carbon dioxide (CO_2) cycle, are promising technologies currently being investigated by the international research community for concentrated solar thermal power generation (CSP). These power cycles have the potential to increase the overall efficiency of the power plant compared to state-of-the-art steam Rankine cycles and thereby reduce the levelized cost of electricity (LCOE) of CSP.Â Therefore, novel thermal energy storage materials and processes are required to successfully couple concentrated solar power collectors with high temperature power cycles.

In this context, the $CaCO_3$/CaO thermochemical cycle is an attractive process for thermal energy storage that can provide dispatchable concentrated solar thermal energy at temperatures between 650â€“980Â°C, appropriate to use with high temperature power cycles. Current research efforts focus on material development and solar reactors. However, there is a need for investigation into the field of system integration to support and verify the viability of this concept.

In this work, we evaluate the performance of a $CaCO_3$/CaO system coupled to a supercritical CO_2 power cycle for continuous electricity production. This investigation has identified high-energy demanding subsystems that require minimization to improve the overall efficiency of the process. The effect of the operational conditions (temperature and pressure) in the calcination receiver and carbonation reactor on overall efficiency are also investigated.

465112 Long Term Stability and Operational Limits of Lithium Sulphate for Thermal Energy Storage

Abstract:

This work explores the experimental feasibility of lithium sulphate as a thermal energy storage material for concentrated solar power (CSP) technologies. This salt has a crystalline phase change between cubic and monoclinic at temperatures around 576°C. By operating with a minimum cycle temperature, both sensible and latent thermal energy can be utilised. The operating temperatures of the proposed system are appropriate for current CSP technologies based on subcritical steam Rankine or supercritical CO_2 power cycles.

Current phase change materials (PCMs) under investigation utilize solid-liquid transformations which can achieve a higher storage performance than sensible storage materials within a low temperature difference. However, these PCMs suffer from some technical problems such as supercooling, corrosion and volume expansion. Such disadvantages can be overcome when solid-solid PCM is used as the storage material as it is less corrosive and undergoes a much smaller volume change.

In this work, the long term performance and stability of lithium sulphate has been evaluated. This involved long term experiments using 55 g of salt in an air furnace, with characterization of material properties before and after cycling using TGA-DSC, XRD and SEM. From experiment, it was determined that the highest temperature at which the thermal properties and volume of the salt were unchanged after repeated cycles was equivalent to its melting point (859°C). At temperatures lower than 130°C, the material is prone to hydration reducing the performance through significant changes to volume, density and consequently thermal conductivity.

474794 Thermal Properties and Storage Analysis of Nano Enhanced Organic Phase Change Material (NEOPCM) for Thermal Energy Storage Application

Abstract:

The objective of the present work is to model the thermal energy storage unit using Nano Enhance Organic Phase Change Materials (NEOPCM). NEOPCM is capable of absorbing or releasing the large quantity of latent heat during phase change. This paper investigatesNEOPCM with Lauric acid as a base material and nanoparticles (Copper Oxide, Aluminum Oxide)as supporting materials for the preparation of NEOPCM. In this paper, it was studied experimentally the effect of combinations of nanoparticles CuO and Al_2O_3 with different mass fractions to the base PCM, Lauric acid. To ensure the uniform dispersion a field emission scanning electron microscope (FESEM) has been used. In order to determine the changes in the compositional/functional groups during the cycle test, Fourier Transform Infrared (FT-IR) spectroscope is used. Crystalloid phase of NEOPCMs is investigated with an x-ray test conducted by an x-ray diffractometer (XRD). Thermal reliability is carried out by accelerated thermal cycle tester, phase change temperature and latent heat of fusion are measure by Differential Scanning Calorimeter (DSC).Thermal conductivity of NEOPCM is measure by using a KD-2 pro, which is based on hot wire technique.It was also verified experimentally that there is improvement in thermal conductivity and storage capacity of NEOPCM.

Keywords: Thermal energy storage, NEOPCM, Lauric acid, differential scanning calorimeter

474795 Thermal properties and heat transfer analysis of Al_2O_3 as a nano particle dispersed in the base mixture of Therminol VP -1 and Water

Abstract:

This paper presents multi-parameter investigation of Nano HTF with Al_2O_3 as a nano particles dispersed in the base material, Therminol VP-1 and water mixture (50/50 vol.%). In this paper, it was studied the effect of different volume fraction of nanoparticle dispersed in to the base HTF, Therminol VP-1 and water mixture. Nano-fluid sample is characterized by Transmission Electron Microscope (TEM) imaging, particle surface charge (zeta potential), particle aggregate size, light absorbance and PH. Dispersed particle aggregate size measure by particle analyzer based on the dynamic light scattering technique. The results show multiple perspective correlations between nano fluid characteristics and thermal properties. Al_2O_3 properties are very close to desired properties of heat transfer fluid properties and it is cheaper. Its thermal conductivity also higher than those of the conventional heat transfer fluids. It was found experimentally that there is significant improvement in thermal conductivity and considerable lowering of viscosity of Nano HTF which is the main requirement of heat transfer fluid for solar thermal applications.

Key words: Therminol VP-1, NEOHTF, Transmission Electron Microscope (TEM), Al_2O_3.

471685 The Solar Fuels Research Program within the Australian Solar Thermal Research Initiative (ASTRI)

Abstract:

The advanced concentrated solar thermal (CST) technologies being developed in the Australian Solar Thermal Research Initiative (ASTRI) can be applied for electricity generation and production of chemical fuels. ASTRI is an eight-year research program supported by the Australian Government, through the Australian Renewable Energy Agency (ARENA). The ASTRI program is a consortium of leading Australian universities and CSIRO, and in close partnership with several international collaborators. In particular, this solar fuels research program within ASTRI aims at demonstrating production of liquid fuels to increase the share of CST in Australiaâ€™s energy supply and lower greenhouse gases emissions.

With a vast array of possible processes and parameters involved in the conversion of solar heat to liquid fuels, a key part of this project has been to identify and prioritise potential pathways. A number chemical pathways, using various feedstock, are being developed and analysed. These include low cost fossil fuels, biomass, CO_2 and H_2O. Technologies being developed to support these processes are:

- A solar hybridised dual fluidised bed gasifier for processing of dry solid feedstocks, such as lignite, biomass or their mixture;
- A solarised supercritical water gasifier for processing of wet solid feedstocks, such as micro-algae;
- A solar redox reactor for syngas production from H_2O and CO_2 by redox cycling;

The proposed solar fuels technologies is complimented by with the development of processes required to convert the syngas, which produced by the above processes into high-value liquid hydrocarbon fuels. Technologies being

investigated to support the advancement of solar thermal production of liquid fuels are:

- Catalytic materials that reduce the sensitivity of the Fischer-Tropsch (FT) reactor to variability of the quality and quantity of the feedstocks and/or lowering cost for the process
- Advanced Sabatier reaction to utilise the waste stream of CO2 emitted from the proposed solar fuels technologies by targeting enhanced production of higher value products than its conventional use for methane, such as propane and methanol.

The overall aim of this paper is to estimate the cost of the production of a drop-in solar fuel (interchangeable with convention fuels) that can be chosen as realistic target with a goal to produce liquid fuels at a cost well below AUD1.50/L (excise-free at the gate of the plant) in Australia.

459558 Solar Thermochemical CO_2 Splitting Via Redox Cycling with Low-Cost Metal Oxide Nanostructures

Abstract:

Solar Thermochemical CO_2 Splitting via Redox Cycling with Low-cost Metal Oxide Nanostructures

Xiang Gao[1], Peter Kreider[2], Alicia Bayon[3], Thomas Gengenbach[4], Jim Hinkley[3], Wojciech LipiÅ„ski[2], Antonio Tricoli[1]*

[1]Nanotechnology Research Laboratory, Research School of Engineering, The Australian National University, Canberra, ACT 2601, Australia

[2]Solar Thermal Group, Research School of Engineering, The Australian National University, Canberra, ACT 2601, Australia

[3]CSIRO Energy, P.O. Box 330, Newcastle, NSW 2300, Australia

[4]CSIRO Manufacturing, Bayview Avenue, Melbourne, VIC 3168, Australia

michael.gao@anu.edu.au

Solar thermochemical two-step metal oxide redox cycles are viewed as a viable pathway to produce syngas via water

splitting (WS) and carbon dioxide splitting (CDS). Syngas is an important intermediate in the creation of commercial products such as hydrocarbon fuels and ammonia. Using a carbothermal reduction step such as methane partial oxidation (MPO) can alleviate some of the process requirements by decreasing the reduction temperature and increasing the fraction of the oxygen carrier undergoing reaction. Recent studies on metal oxides for MPOâ€"CDS/WS solar thermochemical redox cycles have demonstrated potential improvements to the steady-state solar-to-fuel efficiency. A major challenge is engineering redox materials that can maintain reactivity and stability over many cycles, achieve high non-stoichiometry in the reduction step, and promise large-scale production at low cost.

Here, we present a novel low-cost earth abundant metal oxide nanostructure with as much as ten times higher redox capacity and up to 60% faster reaction kinetics compared to state-of-the-art ceria nano-powders. This composite shows promising long-term stability and maintains a significant fraction of its initially high syngas production rates over multiple MPOâ€"CDS cycles. Structural properties such as crystal phase composition and size, specific surface area, pore volume and particle size distribution were characterised before and after thermochemical cycling. These initial results demonstrate a promising pathway towards development of stable, reactive and low-cost redox material for two-step solar thermochemical fuel production.

460593 Evaluating the Effect of Modeling Variables and Experimental Conditions on Material Development for Solar Thermochemical Water Splitting

Abstract:

While solar energy is the most abundant renewable energy resource, the capture, storage, and distribution of it remains a challenge. Solar thermochemical water splitting (STWS) provides a promising route for efficient utilization of this disperse resource since it allows for use of the entire solar spectrum to convert water to an energy dense fuel, H_2. However, despite a significant number of materials having been examined, an optimal redox material to drive this process has yet to be developed. In this work we focus on developing standardized procedures, both computational and experimental, for assessing new STWS active materials. First, we apply density functional theory (DFT) to assess the effects of a number of variables, including crystal structure, cation disorder, and magnetic ordering, on the predicted solar thermal water splitting behavior of spinel and perovskite metal oxides. Accurate modeling of materials using DFT is particularly challenging for high temperature applications, but we show that all of the variables evaluated can critically impact the predicted STWS behavior and must be modeled accurately in order to predict high-temperature material properties. Experimental results from thermogravimetric analysis and redox cycling in a stagnation flow reactor are presented to validate DFT models. Second, direct comparison of experimental results for new active materials is extremely challenging given the wide variety of water splitting conditions and apparatus configurations used. We evaluate a number of known water splitting metal oxides, including ceria, hercynite (spinel), and SLMA (perovskite), at various experimental conditions in order to allow for a direct comparison and to aid in the effort of developing standard experimental procedures for STWS material development.

466776 Continuous Splitting of CO_2 into CO and O_2 Using an Isothermal Redox Membrane Reactor

Abstract:

Renewable liquid hydrocarbon fuels can be produced with two-step thermochemical redox cycles that convert CO_2 and H_2O with concentrated solar energy into renewable CO, H_2 (syngas) and O_2. Syngas can be further processed into liquid hydrocarbon fuels using commercially available Fischer-Tropsch technology. Typically, solar-driven splitting of CO_2 and H_2O are temperature-swing processes with cyclic production of syngas at lower temperatures and O_2 at higher temperatures. Alternatively, this work develops an isothermal redox membrane reactor that continuously produces O_2 and oxygen vacancies at the outside of a metal oxide membrane. The formed oxygen vacancies migrate along an oxygen vacancy gradient across the membrane. At the inside of the membrane the oxygen vacancies are consumed via oxidation of the membrane with CO_2 or H_2O, yielding continuously CO or H_2, respectively. This work employs thermochemical equilibrium computations to determine the thermodynamic limitations of this concept as a function of process parameters and membrane material. The concept is demonstrated experimentally for CO_2 splitting into CO and O_2 using tubular ceria redox membranes and inert alumina membranes as reference. Steady-state gas production rates are analyzed in-process using gas-chromatography. The membranes are characterized via scanning electron microscopy, X-ray diffraction, and energy dispersive X-ray spectroscopy. Thermodynamic limitations are confirmed experimentally via parametric investigation of the effect of operating temperature, partial pressure of O_2, and metal doping of the ceria membranes on the CO and O_2 production rates.

468650 Experimental Investigation of Methane Reformation Via the Nonstoichiometric Ceria Redox Cycle

Abstract:

Dry reformation of methane via the nonstoichiometric ceria (CeO_2 â€“ $CeO_{2\text{-Î´}}$) redox cycle was theoretically and experimentally examined for converting high-temperature solar process heat to syngas. The aforementioned cycle is conducted through two phases: (1) endothermic reduction of ceria in methane and (2) exothermic oxidation of the reduced ceria with carbon dioxide. In both regimes, chemical equilibrium calculations suggest total selectivity and favorability towards syngas production above 827 Â°C, enabling isothermal operation and providing motivation for the following experimental evaluation. An array of seven xenon short arc lamps provided solar-simulated, high-flux radiation to indirectly irradiate a packed bed of ceria situated within a cavity receiver. The endothermic reduction of ceria in methane was demonstrated at three discrete temperatures: 950, 1035, and 1120 Â°C. Upon completion of reduction, nonstoichiometry (Î´) equaled 0.07, 0.21, and 0.24, yielding methane conversions of 9, 41, and 51 % respectively. Subsequent oxidation was first performed with oxygen to indirectly quantify carbon deposition. This proved to be minimal, but its overall long term effect on cycling stability is still under investigation. Finally, isothermal operation was executed at 1120 °C and the solar-to-fuel conversion efficiency, defined as the ratio of calorific value of syngas (H_2 and CO) produced to the combined input solar radiative energy and the caloric value of methane converted, was measured as 3.71 % average and 4.56 % peak. Results indicate that a packed bed configuration is not optimal for this process and further reactor optimization efforts are underway.

464685 Membrane-Free Photo-Electrochemical Cell Design Involving Flow FieldÂ HydrogenÂ Separation

Abstract:

In order for solar fuel research to make an impact on the future of energy, it needs to transition from materials to devices. In the absence of a breakthrough in photo-absorber materials that fit all the requirements of water splitting, reactor designs that can operate efficiently with sub-optimal (but inexpensive materials) are crucial to realise this.

By applying research into membrane-free electrolysers to photo-electrochemical cells (PEC), the performance of such reactors can, theoretically, surpass many designs in literature. The hydrodynamics of the flowing electrolyte between two parallel plates can be controlled through judicious choice of Reynolds and Pechlet numbers for each of the dissolved product species so that convective forces dominate. This would allow product separation of the dissolved hydrogen and oxygen by fluid flow alone.

If the photo-absorber substrate is transparent (e.g. thin film on FTO, Fluorine-doped Tin Oxide), then PEC flow cells could be stacked perpendicular to the light source. This simple, compact, design would improve the overall solar to hydrogen efficiency as the light that is not captured by the photo-absorber on the first pass, is captured on the second or third thus creating a large photo-electrode surface area per unit volume.

A multi-physics model of the proposed cell arrangement has been developed and solved using MATLAB to implement a finite-difference numerical scheme. Using the model, it is possible to optimise fuel production, overall solar-to-hydrogen efficiency and pumping cost using a cost analysis of the various processes.

The results of the model show that there are multiple benefits of such a reactor design, not least in terms of reduced capital cost due to membrane free operation. It also crucially relaxes two key performance constraints of photo-absorbers, namely a) the photo-voltage required is reduced due to lowered ohmic losses in the electrode gap and b) the removal of the classical compromise between the optical absorption and excitation path length. This reactor design could thus help to realise the potential of sub-optimal semiconductor material such as ultra-thin hematite, which have a very high quantum efficiency but low optical absorbance.

We shall further touch on issues of bubble formation and high pressure operation to highlight elements for potential process intensification, also with respect to concurrent solar energy capture below the 1.23 eV threshold.

Experimental verification of the theoretical/model results will be demonstrated with a single membrane-free flow cell employing stable and inexpensive Fe based materials.

474681 Solar High-Temperature Processes for Solar Fuels Production and Solar Energy Storage Applications

Abstract:

When solar energy is employed as the energy source for the production of the raw materials for the synthesis of fuels, the latter are characterized with the term "solar fuels". In the broad sense this term can contain in addition to "solar hydrogen", synthetic liquid hydrocarbons and alcohols that are produced from reactions between H_2 and CO (syngas) that have originated from solar-aided dissociation processes as well as metal powders obtained by solar thermal reduction of metal oxides. Among the various solar energy variances that can be employed to produce such fuels, the thermochemical route uses solar heat at high temperatures supplied by Concentrated Solar Power (CSP) systems – i.e. special mirror assemblies that track the sun - for performing various high-temperature reactions that produce hydrogen or syngas from transformation of various fossil and non-fossil fuels. Such chemical reactions are steam methane (natural gas) reforming (SMR) or CO_2 (or dry) methane reforming (DMR), (reactions (1) and (2) respectively), gasification of solid carbonaceous materials like coal or biomass, or Water Splitting (WS) to hydrogen and oxygen; this last reaction often in combination with Carbon Dioxide Splitting (CDS) to CO and O_2, to produce syngas. In addition, CSP systems can be employed alternatively to photovoltaics for hydrogen/syngas synthesis indirectly, e.g. by supplying the (solar thermal) electricity for high-temperature electrolysis of steam or of steam/CO_2 mixtures.

The so-called "closed-loop" WS thermochemical cycles (TCs) are a series of consecutive chemical reactions ($\geq$ 2), their "net" sum being the splitting of H_2O to H_2 and O_2, where the maximum-temperature (endothermic) step takes place at a temperature lower than that of the single-step water decomposition chemical reaction. Nevertheless, since they involve a highly endothermic step, they need the input of external energy. The most prominent examples of such solar-aided thermochemical cycles are cycles from the so-called sulphur family (the hybrid sulphur cycle described in reactions (3)-(5) and the sulphur-iodine process) and two-step redox cycles via multivalent metal oxide pairs, as in reactions (6) and (7). Each particular route exhibits advantages as well as drawbacks. An advantage of the latter route vs. the Sulphur-based ones is that the same thermochemical principle can be exploited for CDS as well for the production of CO (reaction (8)) that can be further combined with the H_2 produced via reaction (7) to syngas. On the other hand the high-temperature endothermic step of the sulphur cycles, SO_3 splitting to SO_2 and O_2 requires much lower temperatures than WS/CDS.

$CH_4 + H_2O + (\Delta H) \quad 3\ H_2 + CO$ ...(1)

$CH_4 + CO_2 + (\Delta H) \quad 2\ H_2 + 2CO$ ...(2)

$2\ H_2SO_4\ (l) + (\Delta H) \rightarrow 2\ SO_3\ (g) + 2\ H_2O\ (g)$ ...(3)

$2\ SO_3\ (g) + (\Delta H) \leftrightarrow 2\ SO_2\ (g) + O_2\ (g)$ ...(4)

$2\ SO_2\ (g) + 4\ H_2O\ (l) \rightarrow 2\ H_2SO_4\ (l) + 2\ H_2\ (g)$ (electrolysis) ...(5)

$MeO_{oxidized} + (\Delta H) \rightarrow MeO_{reduced} + \frac{1}{2}\ O_2\ (g)$ ...(6)

$MeO_{reduced} + H_2O\ (g) \rightarrow MeO_{oxidized} + H_2\ (g) + (\Delta H)$ ...(7)

$MeO_{reduced} + CO_2\ (g) \rightarrow MeO_{oxidised} + CO\ (g) + (\Delta H)$ ...(8)

Similar chemistry schemes can be employed in the same solar thermal power plant for the so-called thermochemical storage (TCS) of solar energy. For instance, Sulphur-based cycles can be modified to produce solid Sulphur in next steps of the cycle (disproportionation reaction (9)), which can be then combusted in air to produce high-temperature heat and SO_2 (reaction (10)). Similarly, the heat produced by e.g. a solar receiver during on-sun operation can be used to power an endothermic redox oxide reduction like (6); should this reaction be completely reversible the thermal energy can be recovered completely by the reverse reaction (11) taking place during off-sun operation. The "net" result of such cycles (reactions (3), (4), (5), (9), (10) or (6), (11)) is not the production of a particular chemical but the exploitation of the heat effects of the reactions for the "storage" of solar heat.

$2H_2O(l) + 3SO_2(g) \rightarrow 2H_2SO_4(aq) + S(s)$ (disproportionation) ...(9)

$S(l) + O_2(g) \rightarrow SO_2(g)$ ...(10)

$MeO_{reduced} + O_2\ (g) + (N_2\ (g)) \rightarrow MeO_{ox} + (N_2\ (g)) + (\Delta H)$ ...(11)

Large scale production of solar fuels will require the economies of scale offered by heliostat fields with central tower receivers that can comfortably generate solar thermal fluxes in the MW capacity, although the deployment of such fields to date has been limited to electricity generation. For the efficient design and operation of solar aided-reactors for such processes, concepts from "traditional" chemical reactor engineering should be combined with ways to achieve efficient heating of the reactor via concentrated solar irradiation. Among the endothermic steps of the reaction schemes above, reforming and SO_3 splitting are catalytic reactions of gases on a solid catalyst surface. In contrast redox oxide-based cycles involve gas-solid reactions where the solid reactant (metal oxide) is not a "catalyst" present in much smaller quantities than those of the gaseous reactants, but a reactant itself, with non-negligible mass, that not only has to be heated to the reaction temperature but gets progressively depleted during the course of the reaction, having to be replenished. Therefore it either has to be fed constantly into the reactor if the reactions are to be performed in a continuous-mode, or alternatively, practical ways have to be "invented" for its' in-situ regeneration imposing rather challenging reactor operation conditions.

All processes described above, reforming, splitting and TCS, involve not only gas-solid chemical contact and reaction but heat exchange between either the solar receiver and the heat transfer fluid (SMR/DMR, WS/CDS) or the heat transfer fluid and the storage medium (TCS). The major technical challenge for these applications lies in the proper design and operation of reforming/splitting/storage reactors that will operate simultaneously and efficiently as heat exchangers. The two main research tasks are the improvement of solar interfaces and integrated heat recovery schemes in the one hand and solving the main materials-related issues and providing the right functional materials at reasonable costs in the other hand. The present analysis gives an overview on the recent developments and state-of-the art of those concepts at DLR as well as at international level, emphasizing on the most important performance parameters involved and the commonalities that can lead to convergence of reactor concepts employed up until now in different applications. Finally, the current outlook on further potential and necessary developments in the areas of solar fuels production and the use of chemical reactions to store solar heat is discussed.

475638 Redox Active Metal Oxide-Based Thermochemical Processes for Producing Solar Fuels and Storing Thermal EnergyÂ (Invited Talk)

Abstract:

Reversing combustion and recycling carbon dioxide and water back to liquid hydrocarbons is an attractive option for storing solar energy and mitigating the growth of atmospheric CO_2 concentrations. For any such process, high solar-to-fuel efficiency and material availability is critical for large scale viability and favorable economics. Redox active metal oxide-based thermochemical approaches for solar-to-fuel have the potential to be highly efficient as they avoid inherent limitations of photosynthesis and sidestep the solar-to-electric conversion necessary to drive electrolytic reactions. This presentation highlights the challenges and progress from multidisciplinary and international efforts that has been progressing down a technical path for systems, novel reactors, and materials design and discovery for making hydrogen, liquid hydrocarbon fuels, and storing thermal energy from concentrated sunlight based on a two-step metal oxide redox cycle motivated to address the dual challenges posed by the strategic and economic importance of petroleum and the increasing concentration of atmospheric carbon dioxide.

454582 Doped Ceria for Solar Thermal Water Splitting: What Works, What Doesn't, and How to Improve It

Abstract:

Solar thermal water splitting (STWS) has the potential to achieve high solar to hydrogen efficiencies because of its use of the entire solar spectrum to split water. Two-step STWS uses a metal oxide oxygen carrier to couple two reduction/oxidation half reactions, one of which generates O_2 and the other which splits water to generate H_2. However, an optimal material which facilitates these reactions has not been identified. Among the candidate materials that are known to be capable of driving STWS, CeO_2 is considered the most promising. However it requires very high reduction temperatures (≥1500°C) to achieve acceptable H_2 production capacities. Many researchers have tried to dope ceria to bring down the reduction temperature, to varying degrees of success. In this work we use *ab initio* calculations based on density functional theory to develop a fundamental understand and explanation for why only Hf, and Zr have "improved" the STWS abilities of ceria. Based on this understanding we propose a novel set of dopant combinations and modifications strategies that, based on DFT results, improve the hydrogen production capacity of ceria at lower temperatures and/or result in higher H_2O conversion extents during the oxidation reaction.

454960 Stabilizing SiC for Solar Thermal Water Splitting Applications

Abstract:

Stabilizing SiC for Solar Thermal Water Splitting Applications

Amanda Hoskins[1], Aidan Coffey[1], Charles B. Musgrave[1,2], Alan W. Weimer[1]

[1]Department of Chemical and Biological Engineering, University of Colorado, Boulder, CO 80303, USA.

[2]Department of Chemistry and Biochemistry, University of Colorado, Boulder, CO 80309, USA.

Solar thermal water splitting (STWS) has been proposed as a means for hydrogen production which has the benefit that both the energy source and the fuel are renewable and readily available. However, considerable technological challenges must first be addressed in regards to hydrogen production, storage and transport in order for hydrogen to become an economically viable fuel on a large enough scale to significantly contribute to future energy demands. A major limiting factor to the viability of this technology is the lack of high temperature materials able to withstand the extreme chemical and physical environment. Specifically, materials that can be used to fabricate efficient heat exchangers for steam/steam heat recuperation will significantly impact the progress of this technology.

Silicon Carbide (SiC) is an ideal material for high-temperature heat exchange due to its stability at much higher temperatures than the current state-of-the-art material (Inconel) which cannot be used above ~1000°C. Using SiC for heat exchangers involves challenges related to the lifetime of SiC in an oxidative environment. Silicon carbide degrades in water-rich environments through the formation and subsequent degradation of a silica (SiO_2) layer at the exposed surface. In a dry environment, the silica has a low permeability to oxygen, however the presence of water, especially at high temperatures, increases the rate of oxidation, causing cracking and volatilization of the silicon from the structure. We propose that the application of nanostructured films grown with atomic layer deposition (ALD) will significantly improve the oxidation resistance of SiC in extreme environments. We have targeted a variety of high temperature coating materials chosen based on thermal properties, structural characteristics, and stability in oxidative environments. We have grown ALD layers of our desired materials on SiC particles using a fluidized bed particle ALD reactor. We have characterized all films using energy dispersive X-ray spectroscopy (EDS) mapping, scanning electron microscopy (SEM), transmission electron microscopy (TEM), and inductively coupled plasma mass spectrometry (ICP). Using thermogravimetric analysis, it has been shown that these coatings improve the oxidation resistance of SiC.

We have applied density functional theory (DFT) to model the performance of these coatings in an oxidative environment. Computational modeling of the chosen coating material properties, as well as oxygen diffusion through these materials can identify promising coatings to further extend the lifetime of SiC. We expect to use this understanding to develop more advanced stabilizing coatings for SiC and the myriad of other applications that require high-temperature oxidation resistant materials. This presentation will focus on both the fundamental DFT calculations and the experimental results for thin ALD films being developed to protect SiC surfaces from water oxidation at temperatures above 1000°C.

460329 Development and Test Operation of a Demonstration Plant for Sulfuric Acid Splitting at the Dlr Concentrating Solar Power Tower Facility

Abstract:

Sulfuric acid splitting is a key step of the hybrid sulfur cycle (HyS) for solar thermochemical hydrogen production. This exothermal reaction can be divided into two steps: firstly, the evaporation of liquid sulfuric acid (H_2SO_4) at about 400 Â°C forming sulfur trioxide (SO_3), and secondly, the decomposition of SO_3 to sulfur dioxide (SO_2) and oxygen (O_2) at 800 â€“ 1000 Â°C. While the first sub-reaction has fast kinetics, the second one is rather slow and requires the introduction of catalysts to achieve sufficient conversion. Since 2004 the concept of a solar receiver-reactor for sulfuric acid splitting has been developed by DLR and operated in its solar furnace during the European projects HYTHEC and HycycleS. In the follow-up European project SOL2HY2, a scale-up of this concept has been designed, developing a solar tower demonstration plant. This demonstrator has a design flow rate of 1Â l/min of sulfuric acid (50Â w%) and consist of four main components arranged in series and connected by Joule heated piping: a 60 kW electrical evaporator for vaporization of the liquid acid, a solar receiver for superheating the SO_3 to about 1000 °C, an adiabatic reactor with a fixed bed of an iron(III) oxide catalyst and a scrubber. The Joule heated evaporator consists of six vertical steel tubes with a siliconized silicon carbide (SiSiC) inner tubes filled with porous SiSiC foam structures for enhanced heat transfer. Liquid acid is injected at the bottom and vaporizes while passing upwards through the pipes. The vapors are collected in a steel manifold and, subsequently, conveyed to the solar receiver with an outer shell also made of steel. Concentrated radiation from the solar field passes through a quartz glass window closing the receiver and heats up a sectioned absorber composed of porous SiSiC foam structures. After superheating in the receiver, the process gas passed through the catalyst bed for an adiabatic reaction forming SO_2. Before neutralization with sodium hydroxide solution in the scrubber, the SO_2 concentration is measured by a customized gas analysis

system via UV/Vis spectroscopy. The pilot plant was constructed and assembled on the research platform of the DLR concentrating solar power tower facility in Juelich, Germany. The layout of the plant was accompanied and supported by thermo-mechanical modelling of the most important components like the evaporator and solar receiver. Initial operation of the demonstrator was performed with air and water as process fluids. During water operation, the solar receiver reached the predicted design parameters achieving a gas outlet temperature of 1000 °C at an absorber front temperature of 1200 °C and a solar power on aperture of 50 kW. In the adiabatic reactor, however, temperatures of only about 400 °C were measured which are too low for SO_3 decomposition. Therefore, the system was modified placing the catalytic fixed bed directly behind the solar absorber in the outlet section of the receiver. In the following test runs, the temperatures of this adiabatic reaction zone were sufficiently high with a minimum temperature in excess of 750 °C below which the catalyst would be deactivated due to sulfate formation. As a result, testing could proceed with sulfuric acid as the feed successfully demonstrating decomposition of SO_3. A detailed analysis of all results of the systematic on-sun test series is given in the present paper.

466891 Modelling of a Continuous Solar Tubular Aerogel Reactor for the Thermochemical Reduction of CeO_2: The Effect of Residence Time, Particle Diameter and Particle Loading

Abstract:

A simulation is developed and implemented to describe a continuously operated and indirectly irradiated solar reactor for the thermochemical reduction of CeO_2 to Ce_2O_3. An aerogel, consisting of N_2 and CeO_2 (particle diameter between 1-20 microns), enters the reactor at 900 K. The aerogel flows upward through a 30 cm long by 2.54 cm diameter tube, made of Tungsten coated with Silicon Carbide at 2500 K. The model aims at calculating the concentration and temperature profiles within the reaction tube at steady state, the conversion of CeO_2 particles and the Ce_2O_3 particles mass flow exiting the reactor.

The radiative transfer model within the reactor is implemented by considering a gray, non-isothermal, absorbing, emitting and scattering media. Scattering is considered to be anisotropic, as described by the Henyey-Greenstein phase function. Optical parameters where calculated from Mie Scattering theory, weighted according the blackbody spectra at 2500 K. The thermochemical reduction of CeO_2 is modeled by considering an endothermic, first order, source term. Temperature dependence follows the Arrhenius equation. Convective heat and mass transport mechanisms disregard any thermo-hydrodynamic effects, and assume instead plug flow. Finally, gas conduction is accounted and CeO_2 particle diffusivity is estimated from the Einstein-Smoluchowsky equation for brownian motion of spherical particles.

Simulations consider axisymmetric fields, i.e. dependent on the axial and radial position within the tube. First, the radiation heat transfer mechanisms where implemented using an implicit hybrid Monte Carlo-Finite Volume scheme, linearizing the fourth order emission term using a Taylor series expansion. On the other hand, the heat and mass convection, diffusion and reaction where implemented using an explicit iterative scheme. The steady state distributions of all combined physical phenomena are obtained by using the false transient-state approximation, while considering an iterative-splitting scheme.

The parameters considered within this study where the average flow velocity, the CeO_2 particle loading and the particle diameter. Simulations are carried out considering particle loadings of 10 and 100 mg/l, particle diameter of 1,5,10 and 20 microns and average flow velocities of 1,5 and 10 mm/s. Obtained results suggest that high particle loadings tend to favour temperature gradients, which hinder the overall reactor performance. Higher gas velocity lowers the conversion, but slightly increases the mass Ce_2O_3 flow rate. Finally, it is interesting to point out, that higher particle diameter, presents low specific extinction coefficients, which allow relatively high radiation heat transfer rates throughout all the reactor, minimizing temperature gradients and yielding overall higher CeO_2 conversion and mass Ce_2O_3 flow rate.

468882 Conceptual Analysis of Process Alternatives for Solar Thermochemical Methanol Production: The Role of Chemical Storage

Abstract:

Conceptual Analysis of Process Alternatives for Solar Thermochemical Methanol Production:

The Role of Chemical Storage

Bruno A. Calfa[1] and Christos T. Maravelias NOTEREF OLE_LINK1 \h * MERGEFORMAT 1

08D0C9EA79F9BACE118C8200AA004BA90B02000000080000000A0000004F004C0045005F004C0049004E004B0031000000

In this talk, we study the production of fuels and chemicals from a renewable source, namely solar energy. In a solar refinery, solar energy is utilized to convert raw materials, such as CO_2 emitted from fossil fuel power plants (and sometimes with H_2O), into value-added products; e.g., methanol and Fischer-Tropsch (F-T) fuels (Herron et al., 2015). While different solar-based technologies have been investigated to carry out the conversion, in this paper, we consider the solar thermochemical splitting of CO_2 and H_2O (see Meier and Steinfeld (2013) and references therein), which has some similarities with Concentrated Solar Power (CSP) technology in the electric power industry.

The viability, based on systems-level analyses, of a solar-driven (thermochemical cycle technology) process to produce methanol and/or F-T fuels from syngas produced from CO_2 and H_2O splitting has recently been evaluated in the literature (Kim et al., 2011; Kim et al., 2012; Kim et al., 2013; Rytter et al., 2016; Falter, Batteiger, and Sizmann, 2016). However, all aforementioned works considered a continuously operated process. However, one fundamental aspect of the production of solar fuels is the inherent *intermittency* of solar energy. Therefore, since the splitting reactions cannot take place when there is not sufficient solar irradiance (i.e., cloudy days or at night), there is a need to consider *chemical storage*.

In this work, we provide a systems-level analysis of three process configurations, a base case and two alternatives, which employ solar-thermal technology for the conversion of CO_2 and H_2O to syngas, which is further upgraded to methanol. We propose and assess the use of chemical storage to cope with the intermittency of solar-based chemical technologies. The base case requires intermediate gas storage, since all streams are treated as continuous from a high-level perspective (typically assumed in the literature).

First we determine the number of compression stages that minimizes the total annualized cost of compression and storage of CO_2/CO and H_2 streams. Material and energy balances followed by an economic analysis indicate that gas compression and storage significantly contribute to the production cost in the base case, which makes it less economically favorable than its alternatives. We perform sensitivity analysis for key process parameters (e.g., chemical conversion of the solar reactors, cost of the solar field and solar reactors, and solar refinery capacity) to investigate their effect on the production cost of methanol.

The two alternative process configurations do not require gas storage, since the syngas production step is operated entirely intermittently. One process alternative uses a reverse water-gas-shift reactor to further convert CO_2 to CO and correct the stoichiometric number (H_2:CO ratio) required for the methanol synthesis step, whereas the other process alternative has a membrane-based CO_2/CO separation that is operated intermittently. Our analyses show that the process alternatives have overall lower total capital investment and operating costs than the base case, whose compression, storage, and amine-based CO_2/CO separation costs represent major components of the production cost.

References

Falter, C.; Batteiger, V.; and Sizmann, A. 2016. Climate Impact and Economic Feasibility of Solar Thermochemical Jet Fuel Production. Environmental Science & Technology. 50(1):470-477.

Herron, J. A.; Kim, J.; Upadhye, A. A.; Huber, G. W.; and Maravelias, C. T. 2015. A General Framework for the Assessment of Solar Fuel Technologies. Energy & Environmental Science. 8(1):126-157.

Kim, J.; Henao, C. A.; Johnson, T. A.; Dedrick, D. E.; Miller, J. E.; Stechel, E. B.; and Maravelias, C. T. 2011. Methanol Production from CO2 Using Solar-Thermal Energy: Process Development and Techno-Economic Analysis. Energy & Environmental Science. 4(9):3122-3132.

Kim, J.; Johnson, T. A.; Miller, J. E.; Stechel, E. B.; and Maravelias, C. T. 2012. Fuel Production from CO_2 Using Solar-Thermal Energy: System Level Analysis. Energy & Environmental Science. 5(9):8417-8429.

Kim, J.; Miller, J. E.; Maravelias, C. T.; and Stechel, E. B. 2013. Comparative Analysis of Environmental Impact of S2P (Sunshine to Petrol) System for Transportation Fuel Production. Applied Energy. 111:1089-1098.

Meier, A., and Steinfeld, A. 2013. Solar Energy. New York, NY: Springer New York. chapter Solar Energy in Thermochemical Processing, 521-552.

Rytter, E.; Sou__kov‡. F; Lundgren, M. K.; Ge, W.; Nannestad, . D.; Venvik, H. J.; and Hillestad, M. 2016. Process Concepts to Produce Syngas for Fischer-Tropsch Fuels by Solar Thermochemical Splitting of Water And/Or CO_2. Fuel Processing Technology. 145:1-8.

[1] Department of Chemical and Biological Engineering. University of Wisconsin-Madison. Madison, WI 53706. USA.

471163 Hercynite Reduction and ITM Membrane Oxygen Removal System for Solar Thermal Water Splitting

Abstract:

Solar Thermochemical Water Splitting Processing (STWS) is a promising technology for producing renewable H_2[1]. A metal oxide, such as hercynite ($FeAl_2O_4$), is first reduced with high temperature concentrated sunlight to produce O_2[2]. Then, the reduced material is exposed to steam to re-oxidize it and produce H_2. Low O_2 partial pressure is required to drive the reduction reaction. An intriguing potentially highly efficient process is the recycle of inert sweep gas at near atmospheric pressure using a high temperature O_2 ionic transport membrane (ITM)[3], since the recycled gas is already at the required high temperature. In this work, we present preliminary experimental results for hercynite reduction using a TGA and separately the removal of O_2 from 1% O_2/inert down to ppm levels using a laboratory high temperature ITM. The impact of separation energy requirements is evaluated for the overall solar to H_2 cycle.

1. Muhich, C.L., et al., *A review and perspective of efficient hydrogen generation via solar thermal water splitting.* WIREs Energy Environ 2016, **5**:p. 261â€"287. doi: 10.1002/wene.174

2. Muhich, C.L., et al., *Predicting the solar thermochemical water splitting ability and reaction mechanism of metal oxides: a case study of the hercynite family of water splitting cycles.* Energy & Environmental Science, 2015. **8**(12): p. 3687-3699.

3. Ehrhart, B.D., et al., *System efficiency for two-step metal oxide solar thermochemical hydrogen production â€" Part 3: Various methods for achieving low oxygen partial pressures in the reduction reaction.* Int. J Hydrogen Energy (online) http://dx.doi.org/10.1016/j.ijhydene.2016.07.106.

474467 Monte Carlo method to calculate the lifetime efficiency of a solar reactor for reduction of zinc oxyde

Abstract:

Monte Carlo method to calculate the lifetime efficiency of a solar reactor for reduction of zinc oxyde

O. Farges, J.J. B´ezian, and M. El Hafi

Universit´e F´ed´erale de Toulouse Midi-Pyr´en´ees, Mines Albi, UMR CNRS 5302, Centre Rapsodee, Campus Jarlard,

f-81013 Albi ct cedex 09, France

Concentrated solar plant are ommonly used to produce electricity through ther- modynamic cycles (Gemasolar Power Plant, Ivanpah Solar Power Plant, ...) and can also be seen as an alternative to fossil fuel-based methods for H2 generation [5]. High temperature solar thermochemical processes make use of concentrated solar energy to realize metal oxydes reduction. These processes can be considered as a solar energy storage : products may be stored and transported (H2 in this case). Among all metal oxydes reduction, the zinc oxyde to zinc (ZnO/Zn) redox reactions are particularly attractive as the first part of a two step water splitting cycle (eqs. (1a) and (1b)).

As presented on fig. 1, a thermochemical reactor *R*, placed at the top of a solar tower receives concentrated solar energy emited from the sun *S* and reflected by heliostats *H*. Thus, the reactor, feeded with zinc oxyde, reaches high temperature, *ie* above 1400 K, to achieve reduction of zinc oxyde and to procude zinc Zn.

$$ZnO \rightarrow Zn + \frac{1}{2}O_2 \quad (1a)$$

$$Zn + H_2O \rightarrow ZnO + H_2 \quad (1b)$$

In order to evaluate the performance of a thermochemical solar plant and to allow an optimally design of both the reactor and the whole facility (heliostat field,

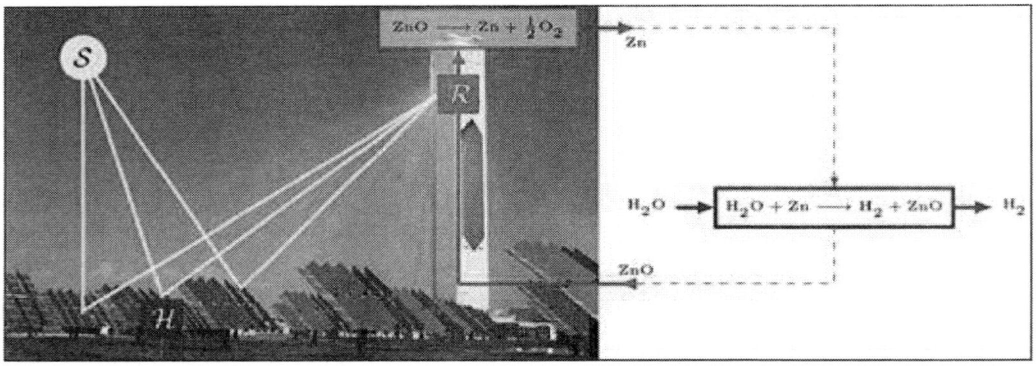

Figure 1: H2 solar production process

tower), numerical modelling of this process is required. We present here a Monte Carlo algorithm that allows the estimation of the solar plant annual productivity which is a non-linear problem. This non-linearity is due to the coupling between photon transport and the metal oxyde reduction reaction. The widespread opinion is currently that "Monte Carlo methods are not generally effective for nonlinear problems mainly because expectations are linear in character"[3]. This difficulty has been overcome by Dauchet who proposed a methodology in his Phd thesis [1].

To evaluate the quantity of interest, *ie* the annual solar-to-chemical conversion rate (ηr), this Monte Carlo algorithm firstly computes the instantaneous power re- ceived at the entrance of the reactor at an instant $\boldsymbol{i}$, $(Pth(\boldsymbol{X}\,|\boldsymbol{i}))$. A Monte Carlo algorithm previously presented in [2] is used as a first step of the proposed algo- rithm to evaluate this quantity. The random variable $Pth(\boldsymbol{X}\,|\boldsymbol{i})$ is the contribution of optical paths $\boldsymbol{X}\,|\boldsymbol{i}$ to the thermal power collected at instant $\boldsymbol{i}$. Then, $(Pth(\boldsymbol{X}\,|\boldsymbol{i}))$, the expectation of this random variable, is the instantaneous thermal power col- lected at moment $\boldsymbol{i}$. A part of the power received by the reactor, the instantaneous useful power $Pu = f(Pth(\boldsymbol{X}\,|\boldsymbol{i}))$, represents the efficient energy used to achieve redox reactions with an instanteneaous reaction rate described by a zero-order Arrhenius-type law, strongly non-linear. Integration over time of the instanta- neous conversion rate leads to the lifetime conversion rate $(\eta r\,((Pth(\boldsymbol{X}\,|\boldsymbol{i}))))\boldsymbol{i}$ as presented eq. (2).

$$\langle \eta_r \rangle = \int_{LifeTime} \frac{P_u(t)}{P_{th}(t)} \quad (2)$$

To treat this non-linearity, the Monte Carlo algorithm is based on a Taylor expansion of $\langle \eta r \rangle$ around $Pth0 > Pth(\boldsymbol{X})$. Independent and equally distributed optical-path random variables $\boldsymbol{Xj} \,|\boldsymbol{I}$ are introduced and the infinite power series is statistically formulated thanks to the discrete random variable J whose realization

j is the order of Taylor expansion. The Taylor expansion is stopped through a Bernoulli process in order to decide whether the algorithm stops or continues without introducing a statistical bias.

Simulations are performed for the 1 MW reactor presented in [4]. A simplified model of thermal losses is applied and a comparison of the nominal performance between results obtained with the Monte Carlo algorithm and experimental results presented in [4] are presented in table 1. The difference between the measured and calculated results are due to the simplified thermal losses model as well as uncertainties of the measurements.

Table 1: Size and performance of a 1 MW reactor

Parameters	Units	Values
Aperture diameter	mm	600
Window diameter	mm	790
Cavity diameter	mm	1600
Cavity length	mm	2400
Conversion rate η_r [4]	%	54
Conversion rate estimated η_r	%	40.2

We have shown in this work that we are able to calculate the efficiency by a Monte Carlo Method, of a solar chemical reactor involving non-linearities between the heat source and the products of the reaction. A futur work could be the improvement of the heat loss model.

References

[1] J. Dauchet. *Analyse radiative des photobioracteurs*. PhD thesis, 2012.

[2] O. Farges, J. B´ezian, H. Bru, M. E. Hafi, R. Fournier, and C. Spiesser. Life- time integration using monte carlo

methods when optimizing the design of concentrated solar power plants. *Solar Energy*, 113(0):57 – 62, 2015.

[3] M. H. Kalos and P. A. Whitlock. *Monte Carlo Methods*. John Wiley & Sons,

2008.

[4] L. Schunk, W. Lipin´ski, and A. Steinfeld. Heat transfer model of a solar receiver-reactor for the thermal dissociation of zno-experimental validation at

10 kw and scale-up to 1 mw. *Chemical Engineering Journal*, 150(23):502 – 508,

2009.

[5] A. Steinfeld. Solar thermochemical production of hydrogen--a review. *Solar*

Energy, 78(5):603-615, 2005.

475639 High Temperature Solar Thermochemical Energy Storage Materials

Abstract:

High temperature solar thermochemical energy storage (STCES) has promise to provide a low cost and high temperature storage solution for solar thermal power.Â However, the current status of STCES is at an early stage of development and substantial research and development is required to realize practical implementation. Two thermochemical energy storage systems with high theoretical energy densities at high temperatures have been investigated. In both the cases, we have been able to develop stable energy storage materials with energy densities in excess of 1200 MJ m^{-3} that can operate at temperatures above 1100 â„ƒ in both cases.

The first system pertains to thermochemical energy storage via decomposition and carbonation of strontium carbonate/strontium oxide in the temperature range of 1150-1235Â°C. Porous structures of pure $SrO/SrCO_3$ were studied via a fractional factorial experimental study to verify the possibility of synthesizing stable undoped $SrO/SrCO_3$ structures using the sacrificial pore formation with graphite (SPFG) method. Results of the experimental study indicate that the SPFG method is not suitable for fabricating porous structures of pure SrO or $SrCO_3$ for high temperature thermochemical energy storage. Cyclic volumetric energy density (including only the chemical reaction contribution) calculated from thermogravimetric measurements decreases substantially to values below 600 MJ m^{-3} upon repeated cycling, demonstrating that the SPFG method not viable. Further investigations revealed that doping $SrO/SrCO_3$ with MgO and BaO improved the cyclic stability significantly. Mixtures of SrO-MgO and SrO-BaO in the molar ratio of 2:3 appear to achieve cyclic stability at 860 ± 5 MJ m^{-3} (500 ± 3 kJ kg^{-1}) and 1408 ± 15 MJ m^{-3} (1001 ± 11 kJ kg^{-1}) respectively over 15 cycles. Pure BaO was also tested for thermochemical energy storage and is found to be more stable with a higher energy density of 1690 ± 24 MJ m^{-3} (1070 ± 15 kJ kg^{-1}).

A second study was performed on transition metal oxide based thermochemical energy storage systems. The purpose of the investigation was to validate the use of magnesium oxide (MgO) as a sintering inhibitor/porous structure stabilizer for high temperature solar thermochemical energy storage (STCES) using reversible redox reactions. Solid solutions of reactive metal oxides in MgO prepared using the solid state reaction method (SSR) were investigated for STCES applications. A new concept of using steam and carbon dioxide to lower the partial pressure of oxygen during thermal reduction was tested and validated. 40 and 33 mole % MnO in MgO, 20 mole % CuO in MgO and 40 mole % Fe_2O_3 in MgO were investigated for high temperature thermochemical energy storage via redox reactions. A composition of 40 mole % MnO in MgO appeared to be the most promising material among those tested here for high temperature thermochemical energy storage. The material can release heat at a temperature greater than 1200°C and has an energy storage capacity of 64212 $kJkg^{-1}$ (122236 MJm^{-3}). The material appeared stable over 20 cycles in a thermogravimetric analyzer and 12 cycles in a tube furnace bench scale reactor and is expected to maintain stability over repeated cycling.

466067 Techno-Economic Assessment of the Polygeneration of Liquid Fuels and Electricity By Solar Hybridised Gasification of Coal

Abstract:

Solar gasification offers the potential to convert a greater portion of the fuel input to syngas than is possible with a conventional gasifier, resulting in a higher production of Fischer-Tropsch liquids (FTL), by displacing the carbonaceous fuel that is otherwise burnt in O_2 to provide process heat. In this hybrid gasification process, concentrated solar thermal energy is used when it is available, while autothermal gasification is used when the incoming solar insolation is insufficient. Thus, the intermittency of the solar resource is accommodated through the use of intermediate storage of syngas and turning down the liquid fuel production rate. This paper presents a techno-economic evaluation of a coal-to-liquids process integrated with a solar hybridised, oxygen blown, atmospheric pressure vortex-flow gasifier (SCTL) located at Daggett, USA. The influence of the solar multiple (SM) and syngas storage capacity (hours) on the plant feasibility was assessed. Here, the SM is the ratio of the heliostat field area to the required heliostat field area that can generate a net annual peak solar thermal energy matching the energy required by the gasification process.

The annual specific FTL output and the specific total energetic output (FTL and electricity) of the SCTL plant relative to the conventional coal-to-liquid (CTL) were found to increase by 12% and 15%, respectively, as the SM was increased to 3. On the other hand, the annual reduction in mine-to-tank (MTT) CO_2 emission from the SCTL relative to the CTL was reduced by 17%. The utilization factor of the FT and power generation plants in the SCTL was significantly influenced by the syngas storage capacity. The utilization factor for the SCTL with no syngas storage was calculated to be 74% compared with 93% for the SCTL with 4 hours syngas storage capacity.

To produce 1500 barrels per day of FTL, the levelised cost of fuel (LCOF) for the CTL was estimated to be USD1.2/L_{2012}. The LCOF for the SCTL with no syngas storage was increased from USD1.3/L_{2012} to USD1.4/L_{2012}, and from USD1.4/L_{2012} to USD1.5/L_{2012} with 4 hours syngas storage capacity, as the SM was increased from 0.5 to 3. Despite the fact that the outputs from the SCTL were higher than that of the CTL, this could not overcome the high cost of the solar components and the syngas storage system. Furthermore, the LCOF for the SCTL is sensitive to several parameters identified in the sensitivity analysis, such as the total plant cost, the assumed cost for carbon capture and sequestration, and the operation and maintenance costs.

452450 Solar Metallothermic Production of Rare Earth Elements

Abstract:

Rare earth elements (REEs) have many applications in metal alloys, catalysts, petroleum refining, glass additives, polishing compounds, and ceramics; they also underpin many critical clean energy technologies such as PV devices, batteries, high-performance magnets, and phosphors for lighting and electronics. The demand for rare earth elements is expected to grow in the coming years, especially in uses related to the energy sector. At present, the extractive processes for REEs and rare earth oxides (REOs) are driven by fossil fuel based energy sources with very high greenhouse gas impacts and resource depletion potential. The separation and reduction stages of the REO/REE production process have the highest energy and material consumption requirements.

Cerium (Ce), lanthanum (La), neodymium (Nd), and praseodymium (Pd) REEs and REOs accounts for 54, 25, 4, and 3% (86% combined) of the global REE supply and usage. Ce, La, Nd, and Pr metals are commercially produced by electrolytic processes where the anhydrous chlorides of the metals are dissolved in molten NaCl, KCl, $CaCl_2$, or other salts. The electrolysis of these rare earth chlorides must occur at approximately 50°C above the melting point of these metals, which are 795, 920, 1021, and 931°C for cerium, lanthanum, neodymium, and praseodymium, respectively.

Energy consumption and environmental concerns, coupled with the fact that the current electrolytic production method must occur at high temperature, provides significant motivation for investigating solar thermal processing as an alternative to conventional REE production. Here, the potential for solar metallothermic production of rare earth elements, especially Ce, La, Nd, and Pr, is critically evaluated using thermodynamic analysis.

473021 Manganese As a Redox Reactant for Sustainable Thermochemical Ammonia Synthesis

Abstract:

Fixed nitrogen, as ammonia (NH_3), is the main ingredient of most fertilizers and its availability is the most common yield-limiting factor in industrial agricultures. Responsible for up to 50% of crop yields, ammonia fertilizers provide the nitrogen necessary to sustain over half of the global population. Annual global ammonia production, currently greater than 140 million metric tons, will need to expand along with the expected increase in demand for fertilizer for food production. Catalytic ammonia synthesis via the Haber-Bosch process, operates at some of the most intense process conditions for chemical production, with temperatures as high as 500°C and pressures reaching 300 atmospheres. The Haber-Bosch process consumes 0.6 tons of natural gas and emits 2.3 tons of carbon dioxide per ton of ammonia produced.

A chemical loop is experimentally demonstrated here using manganese as a cycled redox reactant for ammonia production from nitrogen, water, and methane at atmospheric pressure. Manganese is used to reduce the nitrogen which is then reacted with steam to form ammonia in the presence of a sodium promotor. The corresponding manganese oxide is reduced by a dilute methane stream also producing syngas. The role of sodium hydroxide in the promotion of ammonia liberation is investigated. The thermodynamic predictions of the system are discussed, and the reaction kinetics are evaluated.

The thermochemical cycle appears a promising method to sustainably produce ammonia at near-ambient pressures. Separating the ammonia synthesis into nitrogen reduction and subsequent protonation steps introduces an additional degree of freedom for optimizing the redox reactant and process conditions. Reduced carbon dioxide emissions, and valuable syngas coproduct are advantageous for this novel approach going forward.

462306 ExperimentallyÂ Driven Choices in the Fabrication of a 45 Kw Solar Simulator

Abstract:

Solar-thermal technology has been successfully used to drive biomass gasification, metals reduction and electrical generation in a renewable manner. The development of these applications relies on high-flux solar simulators: assemblies of high power lamps whose irradiance mimics concentrated sunlight. These platforms provide controlled environments for the evaluation solar-thermal technologies at the laboratory scale. To date more than seven high-flux solar simulators are available worldwide for academic studies.In conjunction with Alpha-Omega Power Technologies (LLC), we evaluated choices in high-flux solar simulator construction. This included the effect of different reflector coatings â€“ uncoated, silvered, and aluminized â€“ on lamp efficiency. Additionally, the importance of automated filament positioning was quantified. These results educated the installation of a new, 45 kW, solar simulator at the University of Colorado Boulder.

Carbon Nanotube (CNT) Based Multifunctional Optoelectronic Rods for Efficient Energy Generation and Transport

Sayeeda T. J. Aishee; Evelyn Gonzalez; Erika Hernandez; Glenn Grissom; H. Justin Moore, M. Jasim Uddin*

Department of Chemistry, College of Science, University of Texas at Rio Grande Valley

1 West University Blvd. Brownsville, Texas 78520

*Corresponding Author: jasim.uddin@utrgv.edu, Phone: 956-882-7772

Abstract

To keep up with the world's increasing demand for energy, a clean and perpetual energy source is vital to sustain this planet. This research activity focused on producing a flexible, inter-aligned, thermally stable, highly conductive, carbon based, and ultra-strong photovoltaic (PV) power wire from carbon nanotubes yarn (CNY) functionalized with quantum dots and N719 dye sensitizer. These innovative three dimensional PV wires, with enhanced photoconversion efficiency, have been developed using CNY as the counter electrode (CE) and TiO_2 grafted CNY as a working electrode (WE). The CEs were twisted around the WEs to collect and transmit the photogenerated electrons from the outer circuit. The optimized photovoltaic cell shows a 3.9% photo-energy conversion efficiency.

Introduction

With increasing global awareness for facilitating the energy demand of the world in which climate is constantly changing and pollution negatively affecting many areas, much research has been focused upon finding alternative sources for energy that are both inexhaustible and clean. Among these alternative sources for energy, photovoltaic cells have drawn much attention in the scientific community due to their potential for highly efficient photovoltaic energy conversion [1]. Photovoltaic power wires could revolutionize the traditional solar cells and supply a significant portion of the world's energy consumption. Much work has been done in order to increase the efficiency of the photovoltaic cells; in fact, the energy conversion efficiency of dye-sensitized solar cells has the potential to reach up to 12.1% utilizing platinized transparent conductive oxide anodes [2]. There is a particular interest in creating PV cells that have low internal resistance and low heat transport. It is important to highlight that PV cells are open to textile application, thus greater cell stability is required. Photovoltaic cells constructed from CNY are seen around the world as promising and extraordinary sources for propelling technological advancement in fields involving electronic materials and others, as their ability to convert light to energy is highly favorable and unique [5]. The research focuses on optimizing the energy conversion and improving the mechano-optoelectronic conditions of photovoltaic cells. Enhancing the efficiency of dye-sensitized carbon nanotube yarn shaped photovoltaic cells through the incorporation of MoS_2 is most definitely an innovative project with the potential to affect many fields such as those of textiles, electronics, and engineering in a greatly beneficial manner.

This paper has focused on developing a PV cell from CNY functionalized with molybdenum disulfide and cadmium sulfide as an addition to cell configurations previously reported [1-3, 4]. In order to optimize the efficiency of PV cells, we must first understand its molecular structure of the photoactive phase. The PV microwire was then coated with TiO_2 which acts as a working electrode (WE). Although most PV cells are assembled with a liquid electrolyte, the CNY was used to test the suitability and electronic potential without using any conventional liquid electrolyte. Incorporating CNY around the WE provides a greater sustainability than most traditional dye-sensitized solar cells (DSSC), and it works as a counter electrode. Indeed, the carbon nanotube yarns demonstrate excellent catalytic properties that provide high electronic performance; when the CNY was twisted around the WE, it became the primary conductor of electrons. This power microwire could lead to the widespread use of three dimensional PV wirses in daily life such as clothing, engineering structures, or electronic products.

Experimental Method

In order to develop photovoltaic cells from dye-sensitized CNY functionalized with molybdenum disulfide and attain larger fill factors with an enhanced efficiency thus leading to increased PV energy conversion, the following procedures were undertaken:

Carbon Nanotube Yarn Preparation

The CNY was prepared by removing the polymer from the CNY by soaking it in acetone, and then it was baked at 350°C in a tube furnace with light air flow to facilitate the removal of ash.

Further, the CNY was treated with nitric acid and sonicated to remove any remaining polymer ash.

Working Electrode Treatment

The WE was submerged in Triton-X and rinsed with ethanol. Five yarns were twisted together utilizing their mechanical properties; this process resulted in an assembly of 5 carbon nanotube yarns for the WE. A TiO_2 nano and macro-sol was applied and calcined at 350^0C.

The CNY was then soaked in a solution of $TiCl_4$ and ethanol. Finally, the WE underwent chemical bath deposition. The WE was functionalized with solutions containing quantum dots (molybdenum disulfide and or cadmium disulfide) and dimethyl sulfoxide, taking care to see that these processes did not ruin the integrity of the TiO_2 nano/macro coats. This was confirmed through observation in the atomic force microscope (AFM).

The WE was also placed in mixed solutions of PCBM (Phenyl-C61-butyric acid methyl ester) and P3HT (Poly(3-hexylthiophene-2,5-diyl)) (Scheme 1) , and it was then dried at RT. The PCBM and P_3HT sensitizer based photovoltaic cells facilitate a high efficiency rate in photon to electron conversion. Carrier conversion efficiency (η_c) and energy conversion efficiency (η) of CNY based photovoltaic devices are improved when semiconducting polymers incorporating Bis(1-[3-(methoxycarbonyl)propyl]-1-phenyl)-[6.6]C62 (mixture of isomers), Bis[60] PCBM (PCBM 62) or its functionalized derivatives are used in conjunction with CdS and CdSe. These processes readied the WE to facilitate photoelectron transfer. The WE was then placed in N719 with *tert*-butanol and acetonitrile solution, which is vital to the photosensitive function of the CNY. Any electrodes were kept in the dark prior to assembly, including the WE, which was ready to be assembled into the PV microwires.

Counter Electrode Preparation

The counter electrode was assembled as both single strand CNY and triple strand CNY; this readied the CE for the Pt nanoparticles deposition by sputtering. The sputtering process was done with care to minimize possible damage to the thin film of the CNY.

Electrolyte Preparation

The electrolyte was then prepared with Poly(vinylidene fluoride-hexafluoropropylene) (PVdF-HFP), I_2, and NaI using a chemical bath deposition process while making sure to obtain a smooth homogenous solution. This prepared the electrolyte to be applied to the cells by pipette.

PV Cell Construction

Finally, the PV cell was constructed with a predefined twisting (1 mm pitch distance) process and using conductive silver paste to hold electrodes connection on the slides and connect the ends of the individual electrodes [4]. Approximately thirty PV cells were developed, with half of those having quantum dot incorporated and half of them without it. These CNY cells were compared through analysis of data to confirm the significance of the use of quantum dot in enhancing the efficiency of PV cells with dye-sensitized CNY.

The materials of construction were characterized using Atomic Force Microscope (AFM), the RAMAN Spectroscopy, Infrared (IR) Spectroscopy, and Ultraviolet-Visible Spectroscopy (UV-Vis). The AFM allowed for high-resolution images of the PV cells to be developed and also allowed for sensitive force measurements to be collected. The RAMAN and IR spectroscope yielded much vital data such as exact composition of material, stress/strain state, symmetry and orientation, quality, and amount of material. The IR spectroscope analyzed the infrared light interacting with the PV cell materials by measuring the absorption, emission, and reflection. The UV-Vis yielded data relating to the wavelength and amount of light absorbed by the PV cells. A potentiostat (VERSASTAT 300) was utilized in order to control the current and voltage of the CNY powerwires.

a)

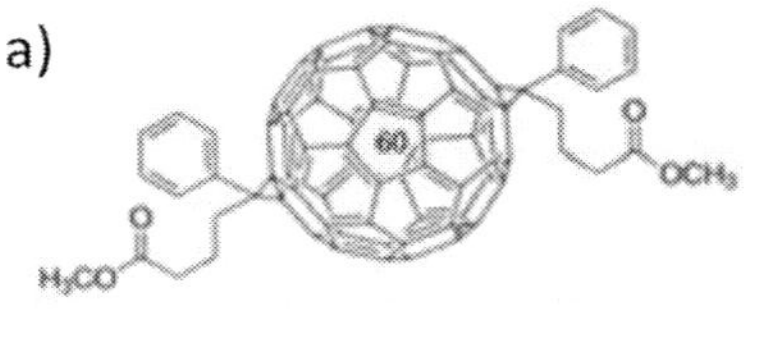

b)

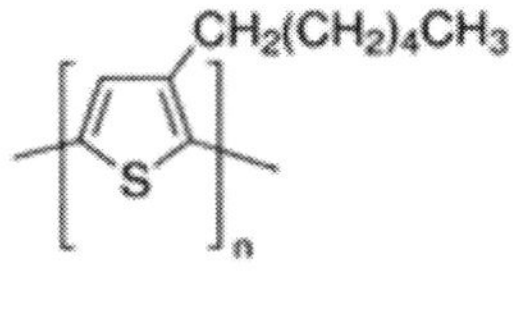

Scheme 1: a) Bis(1-[3-(methoxycarbonyl)propyl]-1-phenyl)-[6.6]C62 (mixture of isomers), Bis[60]PCBM (PCBM 62) and b) Poly(3-hexylthiophene-2,5-diyl) (P3HT).

Results and Discussion

Inventor Computer Aided Design (CAD) Modeling Software was used to better exhibit the construction of the various chemical interlayers of the PV microwires.

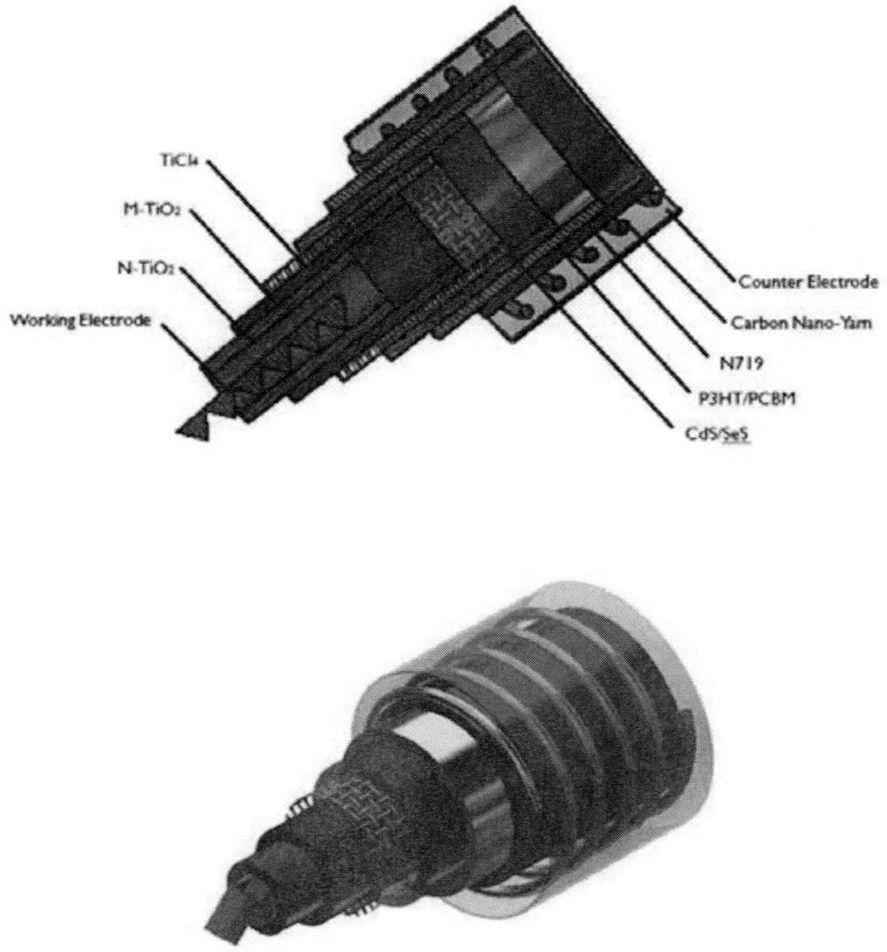

Figure 1a-b: The schematic view of the CNY PV cell.

The PV cells were constructed as shown in Figure 1 and the morphology of the CNY demonstrated with AFM images in Figure 2. The orientation of each individual layer enhanced the energy sensitivity and overall functionality of the cell. As can be seen in figures 2, the CNY clearly exhibits a homogeneous structure without agglomeration. Both images show the underlying carbon nanotube-yarn clearly despite their thick coating and difference in scale. The

CNT yarns are flexible, which enable our WEs and CEs to be twined easily with any of twisting parameters toward the WE and can maintain its structural and chemical properties unchanged after being folded/twisted or tied for multiple times.

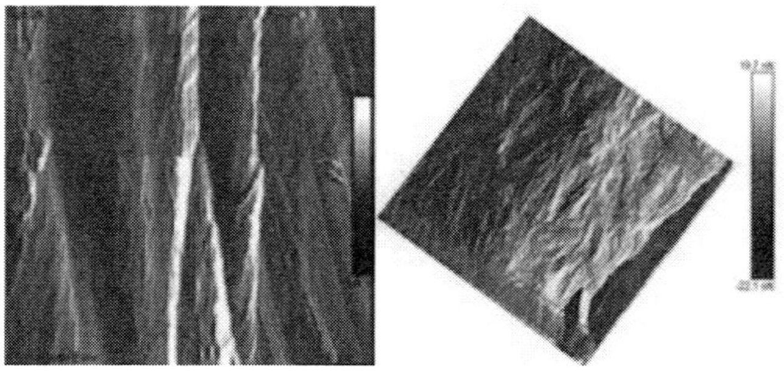

Figures 2 a-b) AFM images of Pristine CNY.

The *I-V* curve exhibits the efficiency of the photovoltaic energy conversion of the cells. This efficiency was calculated by taking the ratio of the maximum electrical power output compared to the solar power input into the PV cell. The fill factor, the measure of quality of the PV cell, was also calculated through the comparison of the maximum power to the theoretical power output at both the short circuit and open circuit voltages. The data obtained from AFM, Raman spectroscopy, and UV-Vis-DRS (Ultra Violet Visible-Diffuse Reflectance spectroscopy) are in-agreement with determining the nanostructured feature of this optoelectronic cells (*the figures are not shown here for the sake of brevity*). The RAMAN spectroscope yielded data that was analyzed in order to determine the structural phase of the WE and CE. The results infer that the titania nanophase are certainly indicative of anatase phase. Finally, data relating to the wavelength and amount of ultraviolet light absorbed by the PV cells was analyzed through the use of the UV-Vis-DRS (figure not shown for the sake of brevity.

The UV–Vis reflectance spectra were carried out at room temperature on a Perkin-Elmer UV–Vis–DRS to detect UV absorption intensities, the location of absorption edges and the quantum size effects of the nanostructured TiO_2 grafted on individual CNTs. The UV–Vis spectra have been obtained in the reflectance mode on the dried samples. This procedure is aimed to avoid systematic errors in evaluating the intensities of the specific UV–Vis absorption bands of the

TiO_2 and CNTs, because, as it is known, different fillings of the intra-crystalline voids by the adsorbed solvent molecules have an important effect on the scattering properties of the supporting phase.

The efficiency of 3D cells is largely dependent upon their fill factor, the measure of quality of the solar cell. We found the fill factor of the cells is ~0.5. Fully functioning photovoltaic cells constructed from dye-sensitized carbon nanotube yarns functionalized with CdS, and MoS_2 (QD) were developed. These PV cells have an enhanced capability for efficient conversion of solar energy, with an increase in efficiency of ~Δ1-2% from that of cells not functionalized with Cds and MoS_2.

Ultimately, the cells constructed from CNY using CdS and MoS_2 had higher fill factors than those without CdS and MoS_2. An increase in fill factor due to the addition of MoS_2 is highly beneficial because it signifies an increase in quality of the PV cells. The energy conversion efficiency of the cells constructed from CNY functionalized with MoS_2 had 3.9% photovoltaic energy conversion efficiency at 0.7 V, as shown in figure 3. Therefore, the PV powerwires were synthesized from CNY using QDs had better efficiency than those without incorporating QDs.

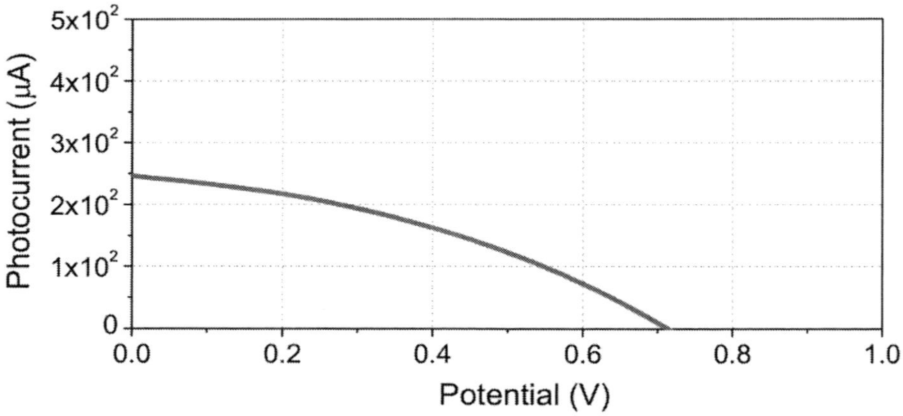

Figure 3: The *I-V* curve of the synthesized cells. The PV cell shows productive usage of power, thus indicating a high efficiency of 3.9% at 0.7 V.

Conclusion

The development of a photovoltaic cell from CNY is advantageous due to the low cost of the materials and the eco-friendly nature of the energy source. The significance of this project exists in the fact that the developed three dimensional PV cells result in much more efficient photovoltaic energy conversion, which makes the future use of photovoltaic cells in everyday life. The application of CdS, MoS_2, and CdSe enhanced the photovoltaic energy conversion productivity of the PV cell. These research outcomes helped maximize the photoenrgy conversion efficiency up to 3.90%. There are many advantages to continuing research on microware structured PV cells, since PV cells are a strong candidate for consideration for future application in a variety of fields such as those involving textiles, electricity, and other forms of engineering.

Acknowledgement:

The authors gratefully acknowledge the research grant (D01_W911SR-14-2-0001-0003) from Department of Defense through MSRDC, Washington, DC 20005.

References

[1] Yan, Jin, Uddin, M. Jasim, Dickens, Tarik J., Daramola, Deborah E., Olawale, David, and Okoli, Okenwa I. 2013. Tailoring the efficiency of 3D wire-shaped photovoltaic cells (WPVCs) by functionalization of solid-liquid interfacial properties. *Phys. Status Solidi* A, 1-7.

[2] Uddin, M.J., Davies B., Dickens, T.J., and Okoli, O.I. 2013. Self-aligned carbon nanotubes yarns (CNY) with efficient optoelectronic interface for microyarn shaped 3D photovoltaic cells. *Solar Energy Materials & Solar Cells* 115, 166-171.

[3] Uddin, M.J., Dickens, T., Yan, J., Chirayath, R., Olawale, D.O., and Okoli, O.I. 2013. Solid state dye-sensitized photovoltaic microwires (DSPMs) with carbon nanotubes yarns as counter electrode: Synthesis and characterization. *Solar Energy Materials & Solar Cells* 108, 65-69.

[4] Uddin, M. Jasim, Grissom, Glenn, Leal, Miguel, Galvez, Veronica, Trad, Tarek, Touhami, Ahmed, Islam, Nazmul, Parsons, Jason, and Moore, H. Justin. 2016. Self-aligned carbon nanotube yarns for multifunctional optoelectronic applications. *Proceedings of the ASME's International Mechanical Engineering Congress and Exposition (IMECE)* 2016-67441, 1-5.

[5] Nonoguchi, Yoshiyuki, Hata, Kenji, and Kawai, Tsuyoshi. 2015. Dispersion of Synthetic MoS_2 Flakes and Their Spontaneous Adsorption on Single-Walled Carbon Nanotubes. *ChemPlusChem* 80, 1158-1163.

464625 Thermal Ageing Characterization of Materials Under Concentrated Solar Power

Abstract:

Concentrating solar power (CSP) technologies have the potential of replacing global natural gas and liquid fuels consumptions by producing solar hydrogen. There are many factors that hinders the competitive low production price of this technology such as degradation of given properties of a solar absorber material over time due to temperature and solar radiation (Brunold *et al.*, 2004). In case of solar absorber materials, the most important factor for aging is degradation of reflectance/absorptance over time and its mechanical integrity (Boubault *et al.*, 2014a). Furthermore, sudden variation of temperature over a short period of time due to variation of solar radiation is another cause of aging (Boubault *et al.*, 2012). It is proposed (Boubault *et al.*, 2014a) that measurement of thermal performance of a material can serve as an ageing criterion (Boubault *et al.*, 2014b), which can be achieved by quantification of its absorption and heat transfer ability. In this work, we will investigate the thermal performance and the optical properties of stainless steel, as a reference material, using a high flux solar simulator and a low power light source. The testing facility consists of a high flux solar simulator which is capable of irradiating Gaussian distributed variable flux (Sarwar *et al.*, 2014). A light collimator was used to homogenize the incident flux. The flux characterization was carried out before and after homogenization using a water cooled Gardon-type circular foil radiometer. The material was exposed to the homogenized flux for given time periods, while we were monitoring continuously the temperature of the material and the heat flux at the back side of the material. The optical properties of the material such as absorption and reflectance were measured and compared before and after exposure to the high concentrated irradiance. Here we present the results of the characterization and the method validation against a total integrated reflectance/transmittance method employing a spectrometer with an integrating sphere.

Acknowledgements

This publication was made possible by a PDRA award [PDRA2-1105-14044] from the Qatar National Research Fund (a member of The Qatar Foundation). The statements made herein are solely the responsibility of the authors.

References

Boubault, A., B. Claudet, O. Faugeroux and G. Olalde, 2014a: Aging of solar absorber materials under highly concentrated solar fluxes. *Solar Energy Materials and Solar Cells*, **123**:211-219.

Boubault, A., B. Claudet, O. Faugeroux and G. Olalde, 2014b: Aging of solar absorber materials under highly concentrated solar fluxes. *Solar Energy Materials and Solar Cells*, **123**:211-219.

Boubault, A., B. Claudet, O. Faugeroux, G. Olalde and J.-J. Serra, 2012: A numerical thermal approach to study the accelerated aging of a solar absorber material. *Solar Energy*, **86**:3153-3167.

Brunold, S., M. Köhl, K. Möller and B. Carlsson, 2004. Chapter 4.4 - Accelerated Indoor Testing. In: M. K. C. J. W. Czanderna (Ed.), *Performance and Durability Assessment* (pp. 175-195). Elsevier: Amsterdam.

Sarwar, J., G. Georgakis, R. LaChance and N. Ozalp, 2014: Description and characterization of an adjustable flux solar simulator for solar thermal, thermochemical and photovoltaic applications. *Solar Energy*, **100**:179-194.

467866 Development and Thermal Test Results of a 10 Kw Porous Media Solar Reactor

Abstract:

Solar volumetric receivers are based on high porosity materials as radiation absorbers, like open cell foams or monolithic honeycombs. These structures allow the volumetric absorption of concentrated solar radiation within them. Solar heat absorbed in the porous media is then transferred to a working fluid passing through it. In this work, a 10 kW volumetric solar reactor has been developed for the reduction of metal oxides. Within the reactor there is a ceramic porous media that supports the metal oxide to be reduced. This structure is directly irradiated with concentrated solar energy in order to reach reaction temperatures above 1300 K.

The developed solar reactor was designed for experimentation with different operational conditions at the IER-UNAM Solar Furnace. The design was accomplished based on detailed ray tracing of the furnace, to obtain the concentrated flux distributions at different planes in the focal zone. The design is based on a cavity with a ceramic conical receiver, and allows for temperature measurements by thermocouples and pyrometers at different spots. Thermal simulations were carried out to determine the reactor body specifications and insulation requirements. The device has been built and preliminary tests have been carried out at the solar. These tests are aimed at evaluating the thermal performance of the system in the absence of chemical reaction. Different porous media that could be used in thermochemical processes have been used in these tests.

468047 Thermal Model of a Solar Porous Media Receiver for a Thermochemical Reactor

Abstract:

Solar receivers containing high porosity materials are a promising option due to higher efficiency compared with tubular receivers. In this system, concentrated solar radiation is absorbed on the porous media in a volumetric manner, then the absorbed heat is transferred to a working fluid when passes through the porous chanels. One critical issue when designing or optimizing volumetric solar absorbers is the knowledge of the temperatures of the solid matrix and fluid. These temperatures are linked with the efficiency and operability of the solar receiver. Hence, this work presents a 3D model of heat transfer and computational fluid dynamics in a porous media receiver subject to concentrated solar flux. This model allow us to analyze the temperature distribution both in the fluid phase and the solid phase, by applying the local thermal non- equilibrium model. The macroscopic model was then used to analyze the impact of geometric properties and operational conditions on the temperature distribution.

474675 Effect of Dust on Straight Tube Receiver of a Parabolic Trough Concentrator

Abstract:

The objective of present investigation is to enhance the performance of parabolic trough collector system. To improve its optical efficiency a dust collector /dirt accumulator has been modelled. This system is being used as parabolic trough based system for power plant application. The experiments were performed to analyze the optical efficiency and optical degradation rate at the roof top of Centre for Energy and Resources Development, Indian Institute of Technology (BHU), Varanasi (25.3176°N,82.9739°E),India. A parametric studies has also been done to estimate the effect of wind flow/air flow, pitch angle, shadow of wall, wind direction ,ambient temperature, solar radiation and collector area for high temperature, high concentration ratio application.

AUTHOR INDEX

AUTHOR INDEX

AUTHOR INDEX

AUTHOR INDEX

AIChE
120 Wall Street, FL 23
New York, NY 10005-4020

ISBN 978-1-5108-3450-7

CURRAN ASSOCIATES INC.
proceedings.com